The Horse in the City

Animals, History, Culture

Harriet Ritvo, Series Editor

The Horse in the City

Living Machines in the Nineteenth Century

CLAY McSHANE AND JOEL A. TARR

The Johns Hopkins University Press
Baltimore

 Published 2007
Printed in the United States of America on acid-free paper
2 4 6 8 9 7 5 3 1

The Johns Hopkins University Press
2715 North Charles Street
Baltimore, Maryland 21218-4363
www.press.jhu.edu

Library of Congress Cataloging-in-Publication Data
McShane, Clay.
The horse in the city : living machines in the nineteenth century /
Clay McShane and Joel A. Tarr.
p. cm.
Includes bibliographical references and index.
ISBN-13: 978-0-8018-8600-3 (hardcover : alk. paper)
ISBN-10: 0-8018-8600-7 (hardcover : alk. paper)
1. Draft horses—United States—History—19th century.
2. Urban animals—United States—History—19th century.
I. Tarr, Joel A. (Joel Arthur), 1934– II. Title.
SF311.3.U6M37 2007
307.76—dc22 2006030213

A catalog record for this book is available from the British Library.

To our children:
Kevin, Susan, and Sharon McShane
Michael, Joanna, Maya, and Ilana Tarr

And to our grandchildren:
Molly, Michael, and Memunatu McShane
Benjamin and Samuel Tarr

CONTENTS

PREFACE

On July 24, 1881, the *New York Times* published an editorial entitled "The Horse in Cities." The editorial noted the horse's indispensability to urban areas but also the high cost at which his services came: "He does earn his living, yet he is a very costly animal." As evidence for the horse's importance, the *Times* listed several items: horses and wagons distributed merchandise throughout the city, horsecars stimulated the development of miles of residential streets, and the desire of the wealthy to "aire" their horses encouraged the creation of city parks.

But, continued the editorial, while the horse was "the most useful animal to man," he also presented many problems. Among the most important was his proneness to disease ("every hair on his body is the name and locus of some equine disorder"), an affliction that had produced the 1872 epizootic. This epizootic, observed the *Times*, had "disorganize[d] everything," and the paper warned that another "extensive horse epidemic" would again deprive the city of fire protection, suspend merchandise traffic, cut off food supplies, and reduce the population to "straits of distress." The horse was problematic not only because he was prone to disease but also because he put riders and pedestrians at "risk of death or maiming" through his "misconduct" and "skittishness." He cost a great deal to feed (he "munches greenbacks when he eats") and had an appetite without limits. And the manure he dropped on the street and the noise and "stone-powder" formed by the pounding of his hoofs on the pavements created health and sanitary problems. In short, concluded the *Times*, although "cities have been made by building around the horse," he should only be regarded as "indispensable" until a better substitute could be found.[1]

This book essentially follows the themes laid out in the *Times* editorial. We explore the use of the horse for hauling freight and passengers, the measures adopted for his regulation and control, his stabling and feeding, his use in leisure activities, his health, and his decline and persistence as an important factor in the urban economy. Our main theme is that the urban horse can be viewed primarily as an animal who was regarded and utilized by a wide variety of urbanites—teamsters, merchants, factory and workshop owners and managers, streetcar drivers and company officials, and even veterinarians—as a *living machine*. This is not to say that these horse users and owners did not also regard the horse as an animal but rather that his animal attributes were evaluated primarily in terms of his ability to contribute constructive work in a variety of contexts. By arguing this, we do not mean to disregard or underestimate the manner in which the horse played a variety of other roles in urban society, such as a companion, an aesthetic object, and a heroic figure for literature, but rather to emphasize that, while these were all important, they pale in significance compared to the dependence of cities upon horses for their vital operations. The significance of the horse as a "living machine" was emphasized for us by the excellent and anonymous reviewer who evaluated our manuscript for the Johns Hopkins University Press. We owe that reviewer a great debt.

The origins of this book actually date back to 1971, when Joel Tarr published "Urban Pollution Many Long Years Ago" in *American Heritage Magazine*. For many years the topic of the horse in the city had been neglected by urban historians until Clay McShane examined the subject in his work on the evolution of the automobile, *Down the Asphalt Path: The Automobile and the American City*, published by Columbia University Press in 1994. In 1997 McShane and Tarr collaborated to prepare "The Centrality of the Horse in the Nineteenth-Century American City," which was published in the second edition of Ray Mohl's urban reader, *The Making of Urban America* (SR Books, 1997). We are indebted to Ray for publishing that article. Since that time the field of animal history has grown vigorously, and several books and articles have been published dealing with different aspects of the horse in society.[2] Readers of this book will find that important subjects relating to horses, such as labor, have not been dealt with in any detail. Because many topics relating to the horse had not been written about previously, we decided not to cover subjects previously covered by historians. Horses had been trotting slowly for too many years, and it was time for this book to leave the stable!

We thank many colleagues who read and commented on portions of the manuscript, including Clay McShane's Northeastern University colleagues Arnold Arluke, Christina Gilmartin, Harvey Green, Anthony Penna, and Anna Suranyi and graduate students and undergraduates at both Northeastern and Carnegie Mellon University who contributed to the project. Colleagues throughout the historical profession also generously provided advice, including Jared Day (Carnegie Mellon), Clifton Hood (Hobart and William Smith), Susan Jones (Colorado–Boulder), James A. McShane (Nebraska–Lincoln), Gijs Mom (Technical University of Eindhoven), Mark Tebeau (Cleveland State), and Terry Sharrer and Roger White (National Museum of American History). Kristen Kurland of Carnegie Mellon gave invaluable help with our maps and charts. We thank Linda Smith Rhoads, editor of *New England Quarterly*, Peter Lyth, editor of the *Journal of Transportation History*, David Goldsmith, editor of the *Journal of Urban History*, and Raymond Mohl, editor of *The Making of Urban America*, second edition (Wilmington, Del.: Scholarly Resources, 1997) for permission to incorporate portions of essays that we wrote for them.

And, of course, there are dozens, if not hundreds of other scholars who have contributed to our ideas, either formally through their publications, as manifested in the notes, or informally through scholarly discourse in real or cyberspace. It is impossible to acknowledge every one of these individuals. That does not limit our debt to them. Bob Brugger, our editor at Hopkins, has been constantly supportive through a long, sometimes tortuous path—as we often noted, the horses were trotting slowly for a time!

The Horse in the City

Thinking about Horses

Humans constructed their understanding and use of horses over millennia. The biologist Jared Diamond's important work notes that horses were perfect domesticable animals with dominance hierarchies, a tolerance for other species, genetic malleability, and herding instincts. In prehistory, the availability of such animals led to the enormous growth and wealth of human populations in areas of the globe where horses existed or could be easily imported. Humans first tamed horses for meat, leather, and manure perhaps as early as 14,000 BCE, even before they were used to carry or pull things. Such beasts of burden allowed significant long-distance trade and economic specialization. By the start of the Christian era, horse-based societies held all the trump cards—more productive agriculture (horses pulled plows, provided fertilizer, and eventually found their way to the dinner table), more long-distance trade, and military supremacy.[1]

The nineteenth-century city represented the climax of human exploitation of horse power. Humans could not have built nor lived in the giant, wealth-generating metropoles that emerged in that century without horses. Horses, too, benefited from the new human ecology. Their populations boomed, and the urban horse, although probably working harder than his rural counterpart, was undoubtedly better fed, better housed, and protected from cruelty. To the extent that it can be determined, the urban horse was also larger and longer lived than were farm animals. Thus, the relationship was symbiotic—horses could not have survived as a species without human intervention, and dense human populations frequently relied on horses. Almost every other species of large grazing mammal disappeared; for example, the original, wild North American horse was unable to defend its territory against smaller predators, including humans. The European horse survived because it found an ecological niche as a partner for humans. In a sense this was co-evolution, not domination.[2]

Mechanizing the Natural

Nineteenth-century business owners valued horses for the profits that they produced. To them, horses were mostly machines, as the French philosopher René Descartes had argued, not living organisms and certainly not moral beings. Robert Bakewell, the famous pioneer of modern breeding, for instance, "sought to discover the animal which was the best machine for turning food into money." *Productive Horse Husbandry*, a frequently reprinted treatise, included a chapter, "The Horse as a Machine," which argued that the best horses were those that produced the most work for the least food (fuel). Another tough-minded author noted that "the horse is looked on as a machine, for sentiment pays no dividend." In the "rational" world of the nineteenth century, people increasingly viewed horses as property or living machines subject to technical refinement, not as sentient beings.[3]

This mechanistic view of animals was most evident when humans compared draft animals to industrial machines. The *Journal of the Franklin Institute* noted in 1833, "The name of horse power has become technical, and is applied to any apparatus by means of which a horse is made to exert his power in propelling machinery." The major aim of the "horse power" machine was to convert the "linear, ambulatory, slightly rhythmic gait of the animal—horse or ox—to the rotary motion required by most machinery," usually through gears. Horses powered machinery in mills and factories, raised and pumped water, sawed wood, drove hoisting devices and construction equipment, and even provided power to drive ferries via paddle wheels and land vehicles via turntables geared to wheels. Numerous museums contain artifacts of such machinery. Obviously, horses hauled things as well.[4]

The problems of engineering with living machines needed to be worked out. How much power could a horse supply? Implicitly this involved knowing how much power an engine or a human could provide, since the point of the comparison was to facilitate engineering decisions about when to adopt human power, when to adopt horse power, and when to adopt steam power. Post-Renaissance European philosophers and scientists had speculated about this question. A scientific approach to comparing the strength of the two species first appeared in 1699 in the initial volume of *Memoires of the French Academy*. The *Memoires* reported a discussion among savants about the horizontal pushing force of a horse and a man. Horses, they speculated, were equal to six or seven men in their power output. Later empirical research would produce results not far different from this.

From the perspective of these French technicians, both humans and horses could be thought of as machines, interchangeable power sources.[5]

When James Watt patented his steam engine in 1775, one of his acts of genius was to create a measure known as "horsepower" and to define it as precisely 33,000 foot-pounds of work per minute. He had derived this figure, still the standard for estimating power, by experimenting with "strong" dray horses because his engines often replaced horses as prime movers in manufacturing processes. One of his first customers was a Nottingham cotton manufacturer who wanted a Watt engine to replace the ten horses that powered his mill. London beer makers, whose horses provided power in their breweries, were also early customers. Purchasers like these needed to know how many horses an engine would replace in order to judge its economic value.[6]

Watt was merely making a crude approximation. Over the next century and a half, engineers and other scientists conducted numerous experiments on how much power a horse generated compared to other prime movers. This research would ultimately lead to the development of the laws of thermodynamics, whose propounders were fully aware that their ideas applied to animals as well as machines.[7] More precise comparisons between the power of horses and of steam engines were possible after French engineer Baron Prony invented the dynamometer in 1821. This was an instrument "to measure force overcoming resistance or producing motion," and many forms of it were developed during the century. One American horse raiser called the dynamometer "the biggest thing that has ever happened in draft horse history. Horses for breeding purposes can now be chosen because of their worth ascertained by actual performance." Obviously, the new device also facilitated comparisons between species or between animals and machines.[8]

As early as 1839, before dynamometers were common, engineer Elwood Morris conducted controlled experiments on horses powering pumps on the Chesapeake and Ohio Canal. Some horses could lift 80 percent more water than others per eight-hour shift. While size was important, it did not explain most of the variation in strength from horse to horse. Morris believed that such factors as wind, lameness, and even "will" explained much of the difference. After eight hours at 3 mph, most horses fatigued. Probably the best-known engineer interested in these matters was Robert Thurston, who founded the American Society of Mechanical Engineers. He is best remembered for his work on steam engines but also wrote about horse power. To him the horse "considered as a machine" depended upon the amount of "stored energy" supplied. Thurston put matters of power into comparative perspective, constructing an elaborate table, broken down by species (oddly, he took no account of variations in ability within each species, although he acknowledged its importance), type of work, and time. For a four-hour shift he concluded humans could move 4,420 foot-pounds per minute, horses 24,780, mules 16,530, and oxen 22,044. In four-hour shifts, humans tired a little more than horses and much less than oxen. By the end of an eight-hour shift, humans tired much more rapidly than did the animals. He also noted, "In the selection and employment of men and animals the engineer is compelled to regard them as machines."[9]

For Thurston, all vertebrate animals, such as the horse or a human, were "prime motor[s]." In this prime motor, he wrote, "the latent forces and energies of a combustible food or a fuel . . . are evolved, transferred and transformed to perform the work of the organism itself, to supply heat to keep it at the temperature necessary for the efficient operation of the machine, and for the performance of external work." Another nineteenth-century authority argued "that animals acting as prime movers have a higher efficiency than any inorganic machines." The efficiency of horses, then, was not just a matter of their strength but also their fuel consumption, initial cost, and maintenance in comparison to other animals, steam engines, or, later, electric motors. For some loads over some distances, horses were more profitable than steam. There was plenty of scientific and technical research on these matters, which is hardly surprising, given the horse's economic importance.[10]

The American Street Railway Association (ASRA) and *Street Railway Journal*, for instance, reported quantified experiments. The Chicago City Street Railway applied dynamometer tests to demonstrate that horses needed to exercise seven times more force to start vehicles than to keep them in motion, an argument for very large animals. In 1890, the ASRA reported, with data from the same company, that horses cost $0.0372 per car mile compared to $0.02371 for electric streetcars. The report undoubtedly sped the disappearance of the horse, since the presidents of virtually every large street railway in the country were at the meeting.[11]

Horses had long pulled omnibuses along city streets, but the spread of the streetcar in the 1850s greatly increased the pulling power of the horse by substituting a smooth rail for a cobblestone street. *American Railroad Journal* pointed out in 1853 that the "power of a horse on the rail is four times as [much as] on the pavement." As a result, horsecars charged considerably less than omnibuses.[12]

Theoretically, the use of steam to power urban transport would have been superior to the use of the living machine. Experiments abounded, especially in the early 1860s, when steam power was still novel and inflation associated with the

Civil War had dramatically increased the cost of horse power. Most attempts to apply steam power to urban transport, however, failed. In 1863, one technical journal cast a discussion of the relative costs of horse and steam power in the context of urban transit and concluded that horses were better. While the *American Railroad Journal* carried numerous articles in the 1860s on light steam engines on street railways, urban residents found the environmental problems of steamers (high-speed accidents, smoke and steam exhaust) to be worse than the problems associated with living machines. People may have held some ambivalence about animals in their neighborhoods, but they feared inorganic machines more. Eventually, for example, a New York court ruled that "the running of the [steam] cars may be regarded as a public nuisance."[13]

City councils usually banned light steam engines, called *dummies* because they were designed to avoid the appearance of locomotives, for environmental reasons, except perhaps on isolated rights of way to new suburbs. Only emergencies allowed exceptions to the rules against steam vehicles. Boston, for example, adopted some steam-pulled fire engines after horse-pulled ones proved inadequate during a conflagration that coincided with an 1872 epizootic (an epizootic is an epidemic among animals), and New York's aldermen authorized steam dummies for streetcars for thirty days during the same epizootic.[14]

Regulation alone, however, cannot explain the failure of steam streetcars. It is reasonable to believe that the entrepreneurs who had the political clout to obtain horse railway franchises also had the ability to overcome anti-steamer environmental regulations, but the higher cost of light steamers was a harder obstacle to overcome. A close analysis in 1860 had claimed that a steam engine cost 42 percent more per day to operate than did horses. In 1866 at least eight New York street railway companies were experimenting with light steam-powered vehicles, but all had given up on costly steam engines by 1870. The living machine had triumphed over the purely mechanical.[15]

The horsecar also looked like a technology capable of enormous improvement through refinements in vehicles and horses. Scientific analysis of horses themselves and not just vehicles became quite sophisticated. Between 1860 and 1900, the *United States Index Catalog of the Surgeon General* listed nearly a hundred works on animal fatigue, a novel category and a sign of the growing attention being paid to matters of fueling an organic power source. Thurston had analyzed metabolic efficiency as well, noting that well-fed animals generated about five calories of heat per gram of oxygen inhaled, making them more efficient as prime movers than steam engines. He compared the oat-fueled to the coal-fueled: "Apart from intelligence and will . . . it [the former] is a self-contained prime mover." Mechanical analysis extended to other elements of horse use. One horsecar operator, for instance, described horses' legs as "weak parts" in a purely mechanical system. Shooting lame horses was commonplace, just as one might junk a worn-out engine. And the dead horse would frequently be recycled at an animal rendering plant, the organic equivalent of a scrap iron processor. Street railways depreciated their horses over five years and then sold them, even if healthy, since statistical analysis showed that lameness occurred more frequently after such heavy work. Lameness, more than fatigue, may have determined the working shifts and even the working life of horses that pounded urban pavements.[16]

The exact nature of a horse's gait became the subject of research, most famously in Edward M. Muybridge's widely publicized 1872 series of photos conclusively demonstrating that galloping horses had all four feet off the ground simultaneously. Trial and error had already shown that trotting was the best gait for pulling carriages. Thurston reported on European tests suggesting that, at 2.5 miles per hour, the optimum speed, horses could pull a heavy load seven times farther than at 10 miles an hour. Thus galloping, already banned in cities for safety reasons, proved economically inefficient as well.[17]

The living machine had attachments developed for the urban setting in addition to the traditional harness, notably blinders and padded, cleated horseshoes. City horses, unlike their country cousins, worked in the cold and rain, so stable manuals written after 1900 always recommended blankets. Wells Fargo, for one, made them mandatory. When straw hats became fashionable for humans in the early twentieth century, they became fashionable for horses as well, ostensibly to prevent sunstroke.[18]

Owners of large herds of horses, such as street railways, applied other "progressive" management techniques to their horses. Streetcar companies replaced horseshoes on schedule, not just when broken, as a form of preventive maintenance. Street railway companies restricted the equine workday to just five hours, far less than human workers. This practice reflected the results of early trial-and-error testing. The constant starts, stops, and heavy loads of street railway service seemed to increase lameness dramatically in animals that worked longer than that. Such practices were always defended in economic, not humane terms. W. J. Gordon's 1893 survey of London draft animals showed that, for other businesses, shifts varied from six to sixteen hours, depending on route, load, downtime, and frequency of stops. Helpers could be added. The Providence street railway kept twenty horses in a small stable at the foot of Constitution Hill to help pull its vehicles up the hill. Stable boys, in training to become drivers, rode the horses back down the steep slope. In some industrial applications horses who turned treadmills only worked three-hour shifts. These practices were remarkably similar from city to city, even country to country, spread through the trade press, conferences, and shifting managers.[19] Finally, the same technicians compared the potential work of horses to men. This comparison eliminated some human work: porters virtually disappeared from cities, and before 1870 horses began to pull fire engines and to tow rotating street sweepers.[20]

In summary, urban owners treated horses mostly as machines, but they also knew that their biological nature produced varying traits that differed from horse to horse. Breeding and castration represented a direct manipulation of the natural. Owners also developed a variety of mechanical attachments, harness, vehicles, improved horseshoes, and blinders to deal with this biological heritage. Although they would not have characterized it this way, they recognized that horses had gender, feelings, intellectual ability, and deeply ingrained behavioral patterns, an evolutionary legacy. Humans also provided horses with more suitable environmental circumstances, notably smooth, easy-to-grip streets and healthier stables.[21]

Roots

It took humans millennia to develop the horse as a power source, but the horse proved to be an evolving, malleable technology. Some background on the history of domestication explains much. With few exceptions it is possible to control only animals that have herding and following instincts, such as horses, dogs, or cattle (or, for that matter, human slaves). Castration, or gelding, a practice borrowed from human slavery, made taming stallions easier. In a sense, humans substituted themselves for the herd leaders, turning equine herding instincts against the animals. Taming horses was not enough; humans domesticated them, too (that is, controlled their reproduction).[22]

Taming horses for travel depended on control technologies, beginning with the invention of the bit, a metal brace placed in a gap between a horse's teeth. When drivers pull a rein connected to the end of a bit, they direct horses to turn in the direction of the pull. When they pull both reins, the best way for a horse to halt the pain is to stop. The whip was another harsh control mechanism. Presumably humans in the ancient world had more positive taming methods as well. Horse owners soon discovered that horses tended to bond with individual riders, since horses can recognize individual voices and intonations. Food, even sweets, might be a reward and deprivation a punishment.[23]

Horses carried an enormous symbolic load as well, in part because they represented the human triumph over nature, in part because of their military applications. They became the tool of warrior castes, which had the time to master them and the resources to feed them. All ancient Indo-European cultures valued the horse, which they rode into domination. Since the days of Charlemagne, Western artists have always depicted generals and kings on their horses. The horse became a symbol of status and the social order. Horses were construed as more than just animals.[24]

Two-wheel carts had been common since Roman times, where paved roads existed, but four-wheeled vehicles were extremely rare. Until well into the nineteenth century, intercity land freight was much more likely to travel by packhorse than by wheeled vehicles. Much of the increase in horse use in the nineteenth century can be attributed to better vehicles, but that topic is so complex, we have paid it only cursory attention.[25]

In medieval (and, for that matter, contemporary) legal theory, horses, like other animals, are merely chattels, subject to a great extent to their owners' whims, so one living being could own another (the same applied to human slavery). The medieval love of blood sports like bear baiting and bull fighting suggests a cruel attitude toward their chattels. There were ambiguities. Some Christians emphasized that animals were also part of God's creation and anthropomorphized them. Saint Francis spoke to the birds. Talking donkeys also appear in saintly legends. On the secular side, Chaucer wrote tales humanizing animals. The relative rarity of gelding, the naming of horses, the custom of courtroom prosecutions of animals (mostly for killing humans)—a practice that existed even in colonial New England—and the taboo on eating horsemeat in much of western and southern Europe suggest a belief that somehow animals had a kindred nature.[26]

New ideas about animals that appeared in the early modern period would have great influence in North America. There were some sixteenth-century theriophiles (animal lovers) who argued for animal intelligence, but the French philosopher René Descartes laid out the dominant thought of the new era. Descartes believed that animals were purely mechanical, incapable of reason or emotion, even pain. In general, Enlightenment thinkers saw nature (including animals) as a resource for humans, not as something intrinsically valuable.[27]

The practice of animal husbandry in colonial North America was quite poor. Horses were rare, and tilling with hoes by hand took the place of ploughing, the most common use of the horse. Breeding was a hit-or-miss affair, and animals ate almost exclusively by grazing, a problem during North American winters. American legislatures, greatly influenced by Enlightenment ideas, evidently thought about animals mostly as agricultural tools (although both Quaker Pennsylvania and Puritan Massachusetts passed anti-cruelty laws, and there was a common law prosecution for beating a horse as early as 1788). Human porters, to judge from iconographic evidence, carried most urban freight.[28]

Victorian thinkers, whether in the United States or Europe, were somewhat more ambivalent. Many accepted Darwinian notions about the survival of the fittest. For them, hegemony over animals was the norm. Other Victorians were vitalists, believing that animals and even plants had a life force that science could not understand. Thus Thoreau wrote, "Every creature is better alive than dead."[29] Additionally, he said, "It must be confessed that horses at present work too exclusively for men, rarely men for horses."[30] Even Darwin held a respect for natural beings; he belonged to humane societies and became a leading antivivisectionist.[31] Charles Dickens ridiculed a schoolboy who defined a horse in purely mechanical terms: "Quadruped. Graminivorous. Forty teeth, namely twenty-four grinders, four eye-teeth, and twelve incisive. Sheds coat in the spring; in marshy countries, sheds hoofs, too. Hoofs hard, but requiring to be shod with iron. Age known by marks in mouth."[32]

A literary genre that saw animals as other than an exploitable commodity began to flourish in the late nineteenth century. Anthropomorphic animals were also a staple of nineteenth-century children's literature, suggesting a perceived kinship, since they are presented as cultural representations of humans. *Black Beauty,* written by Anna Sewell in 1877 and appearing in twelve American editions before 1900, was one of the earliest and most popular works in this mode. Numerous other nineteenth-century authors wrote stories anthropomorphizing animals, usually for juvenile audiences, including Joel Chandler Harris, Jack London, Rudyard Kipling, and Lewis Carroll. This current ran, of course, against the larger theme of exploitation and moderated it because teamsters and owners internalized some of these values, which led to anti-cruelty regulation. Teamsters named and groomed horses and owners entered them in workhorse parades, essentially a form of beauty pageant, all signs that they recognized that their machines were also living organisms.[33]

Creating the Urban Horse: Breeding

The demands of the urban market produced a unique life course for the urban horse. From birth to death, horses were, as one well-known scholar points out, "living commodities with cultural values."[34] Most obviously, the new and lucrative urban freight and street railway market, both of which demanded larger horses, changed breeding practices. Before the 1830s, American farmers had raised urban horses in each city's hinterland with little attention to the idea of a breed. Often farmers set aside those animals unable to work, such as the lame and blind, for breeding purposes.[35] They seem to have viewed breeding as a hit-or-miss process. Conestogas and Morgans, the only American animals that might be called breeds, were too light for heavy hauling and varied so widely from animal to animal that the breed name meant almost nothing.

The completion of the American railroad network at midcentury allowed enormous specialization in all agricultural products, including horses. Specialized draft animal breeders appeared in midwestern states where grasses had a high calcium content, which was better for grazing yearlings. There were some subspecialties. Kentucky ranchers raised lighter breeds of horses, and Missouri breeders grew mules. Heavy draft animals probably came from Ohio, Indiana, and especially Illinois. Competition hurt horse producers in traditional breeding areas, less suited for feeding. The Massachusetts State Board of Agriculture noted in 1882 that the commonwealth was finished as a horse producer. The older urban hinterland producers had concentrated on "general purpose" horses that no longer served. The new equine division of labor demanded thoroughbreds for racing, heavy animals for urban hauling, more agile animals for light delivery purposes, and mules for warmer climates. There was even a gender division of labor. Mares remained on farms; in a growth market their reproductive abilities made them too valuable for anything but carrying foals.[36]

America's horse population was nearly as diverse as its human population. The United States had never developed a national breed of draft horse like Shires in England, Clydesdales in Scotland, Westphalians in Germany, or Percherons in France. American farmers, however, were little bound by tradition and quite willing to draw on the European experience. The best example of this technology transfer may be Ohio's State Board of Agriculture. Its secretary for nearly fifty years was J. H. Klippert, the son of German immigrants to Cincinnati. Klippert translated and published the latest reports on breeding from England, France, and Germany during his tenure, and the board thought enough of this approach to send him to Europe on research trips. In any case, importing European horses was a necessity, since American equine workers, like human workers, could not meet growing demand. American breeders imported large numbers of their favored breeds. Immigration was not without its vagaries. Sometimes European sellers fobbed half-breeds or even sterile stallions off on Americans. Trans-Atlantic travel was even harder on equine than human immigrants, with mortality sometimes exceeding 50 percent.[37]

The importation of the French Percheron stallion Louis Napoleon in 1851 triggered the most important wave of immigration. Americans had imported a few stallions from France in both the 1660s and 1840s, but these large horses consumed so much food that the limited freight business available during those periods could not support them. They were absorbed into the general American herd, leaving little trace. When the demand for heavy urban draft horses grew in the mid-nineteenth century, breeders imported more stallions from France than elsewhere. Street railways especially valued Percherons' combination of stamina and pulling power.[38]

The reproductive powers of Percheron stallions were legendary and needed in a society with a rapidly growing demand for heavy draft animals. Percherons were valued not just for siring large numbers of foals but also because they were much more likely to pass on their characteristics. When Louis Napoleon returned to Dupage County, Illinois, after an eastern trip in 1862, four hundred mares were waiting for him, even at his expensive stud fee of ten dollars. A Massachusetts agricultural journal noted that he had served mares as small as nine hundred pounds with "perfect gentleness." The same journal noted that his progeny were "commonplace" in Boston and easily recognizable because they were two hands (eight inches) taller than other horses. Old Bill, another pioneer Percheron, sired over a thousand foals between 1853 and 1874.

Louis Napoleon's colts sold for $600 to $1,000 each as urban draft animals in the 1850s, four to ten times more than existing domestic stock. Prices for very good stallions went even higher. In 1888, a wealthy Wisconsin farmer purchased a famous domestically bred Percheron stallion, Seducteur, for $6,000. Good stallions with French government registration could cost up to $10,000. French breeders were selling 1,500 Percheron stallions a year to Americans by 1880. In the 1880s, more than 40,000 horses came to the United States from Europe each year. When prices for horses crashed during the 1893 depression, farmers got the Treasury Department to impose a tariff on imports, except purebred stallions.[39]

It is easy to ridicule nineteenth-century breeders from the perspective of modern genetics. They probably bred weaknesses in with size; they were obsessed with color, seeking horses of only one color (apparently a notion derived from racial ideas about humans); they believed that they had produced characteristics that seem improbable. One Percheron breeder wrote, explaining the popularity of that breed, "The American has little time to waste on stupid horses."[40] Thinking about such goals could only cloud the most important issue—building a larger living machine. Breeders mostly followed traditional folk genetics, which insisted that only one parent transmitted genes or that the roles of parents were otherwise uneven; that acquired characteristics could be inherited (a Lamarckian notion); that insemination by a mongrel (or horse of a different breed) made a mother's blood impure, affecting subsequent progeny with other fathers (apparently another notion that grew out of traditional human racism); and that accidents to the mother could be passed on to the child (fright, mutilation, etc.). One believed that "nervous owners have fewer foals."[41]

Even J. H. Sanders, the editor of *Breeder's Gazette* and perhaps the most sophisticated breeder of the era (he advocated breed registration before most Americans and claimed to be the inventor of artificial insemination), wrote that the first sire could contaminate the blood of a mare so that his qualities were passed on to subsequent foals with different fathers. He acknowledged that this was a rarity. He added that mental impressions of a dam at conception only rarely affected foals, as with a gray stabled with a bay who had a bay foal when mated to another gray.[42] Note the insertion of the adverb "rarely," a sign that those traditional folk beliefs remained, although in decline. Sanders opposed allowing horses to breed on Sunday.[43]

Still, the breeding system worked, and there were extraordinary increases in equine speed and size in the nineteenth century. Johnstone noted that French Percherons had weighed 1,200 to 1,300 pounds in 1830. The winner at the Ohio State Fair in 1878 weighed 2,015 pounds. He may not even have been the heaviest horse, since the fair judged by conformation and gait, not by pulling contests or weight. *Harper's Monthly* (March 16, 1912) had a photo of a horse belonging to a firm of New York truckmen that "is said to be the largest horse in the world," weighing 2,430 pounds and standing 20.5 hands (7 feet) at the withers. Sixty years earlier the largest urban draft animals would have weighed little more than half that. Equine improvement is more precisely measured with racing trotters because a consistent performance standard is available. Between 1806 and 1903, the one-mile U.S. trotting record dropped from 2 minutes, 59 seconds to 2.00. Even with a far more accurate understanding of genetics, that time dropped only a further ten seconds or so in the twentieth century. These nineteenth-century breeders were successful to the point that Japan imported American, not European, stallions when it began to modernize its draft horse herd in 1877. After 1895 the United States became a net exporter of draft horses to Europe. During the Boer War, British purchasing agents decided that U.S. heavy horses were better than domestic ones and bought extensively in the United States. In 1897, a Colonel Holloway of Alexis, Illinois, exported Prince Sturdy and Prince Cedric, two Clydesdale stallions bred in the United States, to Britain.[44]

Americans experimented with other breeds as well. Belgians, often stronger than Percherons, were the next most popular breed, but they were slower, uglier, and relatively short-lived. They began to displace Percherons in the 1890s, since they were better feeders. English Shire horses were stronger yet but even shorter lived. Like other British breeds, they had hairy legs (feathered fetlocks in horse talk) that made them prone to leg infections in North American conditions because the constant freeze/thaw winter cycle froze to their legs the slush or mud from poorly paved streets. Heavy feathering also made grooming more tedious. American breeders had imported some Shires from Ontario as early as the 1850s. Some Clydesdales migrated from Scotland in the 1850s, and in 1875 a Chicago auctioneer listed seventy-five Clydesdales in his catalog. Clydesdale imports picked up markedly in the 1880s. They had a high-stepping gait and appeared taller than they were because of their long, erect necks, appearances highly esteemed by nineteenth-century horse fanciers. Businesses that delivered their own goods preferred handsome horses, like Clydesdales. Sorry-looking nags were not good advertisements for their businesses and wound up with industrial deliveries or low-prestige peddlers. Clydesdales had reputations for being temperamental, unhealthy (feathered fetlocks were an issue here, too), and short-lived. Britain's horse market was less well regulated than the French market, so dealers who exported unsound horses lowered the reputations of Clydesdales, Shires, and a less numerous British breed, the Suffolk Punch, diminishing sales from that nation.[45]

Mules

One other draft animal played a significant role in the urban scene—the mule. In Roman times, humans began to breed mules, the sterile offspring of a male donkey and a female horse. Mules may be the best demonstration of the mutability of equine DNA. The breed (or species—it is hard to figure out what to call a group of animals that doesn't reproduce) is the purely artificial creation of humans. They never existed in nature. While not genetically engineered in the modern sense of the word, they certainly come close. Their sterility also reduces antagonistic behavior and facilitates bonding with humans. We have chosen to treat them as a breed of horse, as did nineteenth-century owners of draft animals, who saw mules as just another option, similar to the choice among various breeds of horses.

Mules were more numerous, by millions, than any specific breed of horse in the United States. Their numbers doubled in the 1850s and continued to grow thereafter, especially on farms. As late as 1922 there were more than five million mules in the nation, mostly in the rural South. Mules were cheap to buy, long-lived (often doubling the working life of horses), and very good feeders. Except in a few, warm weather cities (mules supposedly handled heat better and disliked snow), urban teamsters avoided them, in part because they were weaker than horses, rarely reaching 14 hands (56 inches), in part because their truculence was legendary and probably was worse in traffic. One contemporary described them as "vicious, stubborn, and slow." This may have reflected the fact that mules, unlike horses, would not work if underfed or beaten. C. G. Goodrich of the Minneapolis Street Railways noted that he employed 560 mules, averaging 14 hands and 800–900 pounds. He claimed that they ate only half as much and did twice the work of a horse. His firm gave them up three years later, saying that they were too slow and could not handle snow. New York street railways, as well as those in other cities, such as Pittsburgh and Philadelphia, also employed mules but only during the 1860s, perhaps because the Civil War had increased the price of horses. Northern complaints about the slowness of mules may reflect the unwillingness of these overworked beasts to be rushed into keeping a schedule. One agronomist wrote that "a mule takes better care of himself in the hands of an incompetent driver than a horse does."[46]

Influences

Historians have largely neglected the tremendous influence of the horse, economically as well as culturally, in agriculture, transport, sport, and war, as a companion to humans, a cause of accidents, a prime mover of machines, a source of food, and, of course, a shaper of cities. Rather, they have seen the horse merely as a relic of pre-industrial times. Industrialization and, in general, the modernization of society have been analyzed almost purely from the viewpoint of mechanization. From this perspective it is not surprising that many historians perceive the study of horses as synonymous with amateur nostalgia. Winners write history, or so the aphorism goes, and in the long run the coal-fueled virtually eliminated the oat-fueled. Numerous scholars have written about the triumph of the steam engine in the nineteenth century and ignored the living engine. This view of the horse is wrong-headed—the nineteenth century also marked the triumph of the horse. The horse was a flexible, evolving technology and, like its accomplice, the steam engine, was crucial to the evolution of the modern city. The steam engine actually expanded the role of the horse.[47]

In the early twenty-first century, the U.S. government is quite willing to patent microorganisms (whether that's a wise policy is a different issue). The nineteenth-century marketplace also encouraged the manipulation of living organisms, albeit with a different technique. Perhaps the best measure of the success of the horse and one of the things that triggered our interest is the popular icon of fire horses, noses flaring, galloping to a conflagration pulling a smoking, four-ton, steam-powered, coal-fired pump. It would seem to contradict contemporary experience to have a living entity pull a mechanical one.

Urban historians have long been concerned with the consequences of putting diverse humans in a very densely populated space. We are concerned with a related set of questions, since the horse was far more than just a source of mechanical energy. What happens when you add another species to that space, also at a very high density? How did equine rural-urban migration occur? What was its source and management? What sort of housing did the new migrants have? Were urban horses adequately nourished? What were their working conditions? Why and how did their productivity improve, so that they could avoid mechanical replacement for decades? What happened to their wastes? What happened when they died? Did urban overcrowding of horses create public health issues akin to those faced by humans or pose a danger to humans? Additionally, there are complex questions of interspecies relationships. Who controlled the animals and how? How did the behavior of owners and their hired drivers vary toward their fellow mammals? How was cruelty, a problem even in intraspecies behavior, managed? In general, what were the possibilities and limitations of this labor-intensive power source? Most importantly, how did contemporaries reconcile the-horse-as-a-profitable-power-source with the-horse-as-a-living-organism?

William Cronon, a leading environmental historian, notes that a "rural landscape which omits the city and an urban landscape which omits the country are radically incomplete as portraits of their shared world."[48] Urban horse populations, just like urban human populations, were dependent upon the flow of agricultural products from the hinterland into the city; indeed, they were agricultural products themselves. In this context we will examine the relationship between cities and their hinterlands in the growth and processing of feed crops for urban horses, a relationship that the emergence of the national railroad network dramatically changed. Specifically, how was the output and transfer cityward of draft horses, as well as the hay and oats they required, increased? What role did horse manure play in fertilizing garden farms near the city, and what problems were created when artificial fertilizers replaced manure? How was this aspect of the city-hinterland relationship managed?

Nineteenth-century Americans solved all of these questions in a way that encouraged an enormous growth of the urban horse population. Census and city records suggest that horses were urbanizing more rapidly than people in the third quarter of the nineteenth century, a measure of their indispensability. Boston, not an especially horse-dependent city but one for which there are good records, reflects this pattern. Between 1741, when horses were taxed for the first time, and 1841, there were roughly forty humans for each horse. By 1880 the ratio had dropped to twenty-five. It was back up to forty in 1900, after the electrification of street railways. Over the next ten years, the number of horses grew slightly but did not even come close to keeping up with the growth rate of the human population, and then it dropped dramatically. By 1920 the city had fewer horses than in 1820.[49]

Large nineteenth-century cities averaged roughly one horse for every twenty people, although the ratio of humans to horses varied widely, as reflected in the following examples of numbers of humans per horse in 1900.[50]

Kansas City	7.4
Minneapolis–St. Paul	9.3
Los Angeles	12.7
Denver	14.7
Memphis	17.0
St. Louis	17.5
Buffalo	18.5
San Francisco	20.1
Columbus	20.8
Chicago	22.9
Pittsburgh	23
Cincinnati	23.3
Philadelphia	25.3
New York	26.4

Even where the ratio was high, the total number of horses could be quite large. For example, 130,000 horses worked in Manhattan in 1900, while Chicago had 74,000 horses and Philadelphia 51,000.[51] In 1879 the *New York Times* exaggerated only a little when it noted that "New York must move on wheels, it would be thought—the whole population must drive . . . this is obviously a *Stable* city."[52] Although we've not looked much at western and southern cities, we should note that horse populations were larger in the West. Perhaps the lower population density of western cities made them more horse dependent; perhaps it was cheaper to keep horses. We leave that question to other scholars. These figures do not include animals working in the city but stabled on cheap suburban land or farm animals that slogged into a nearby metropolis every day, carrying fresh produce to urban consumers and often bringing products from the city home. For most of the century, the horse was indispensable as a living machine, to the point that we can label the nineteenth century as the golden age of the horse.

CHAPTER ONE

Markets

The Urban Horse as a Commodity

Nineteenth-century animal owners valued horses almost exclusively for their productive utility. In other words, horses became living machines to be bought and sold like commodities, valued only rarely as natural beings. The willingness of horse owners to end the life of animals that had become even slightly lame—that is, lost their productive value—is a powerful measure of this outlook. Presumably, if horses were companion animals like modern pets or were living in nature, they would have been allowed to graze out their lives to the extent that their disability allowed. The killing of injured but live horses occurred not just because hauling, carrying, and lifting were the only economic functions of horses but also because their carcasses carried considerable economic value, generating income for tanners and renderers, among others. Even the wastes of horses generated income for their human masters, since manure was a valued fertilizer in urban hinterlands. Par excellence, the trade in living beings, used to the extent possible as machines, typified what has been described as the late Victorian definition of civilization: "the necessary, rational management of nature."[1]

Horses also functioned as consumers. Growing hay and oats for urban animals generated income for millions of farmers, while the transport and sale of these agricultural products required additional workers and facilities. The horse economy demanded manufactured goods as well. Horses needed a variety of products, including harnesses, blankets, and shoes. Horses provided jobs for teamsters, blacksmiths and stable hands, and other blue-collar workers. The horse economy also created white-collar jobs for stable supervisors, veterinarians, middlemen in the horse trade, and so on. The reliance upon horses for transportation of people and goods required cities to build new infrastructure around their needs, especially wide streets paved with stone blocks and street rails.

Marketing from Farm to City

Horses moved from farm to city via an exceedingly complex and evolving market process, one that became progressively more specialized throughout the nineteenth century. The most remarkable element of the market for horses was the individuality of transactions. Each horse was sold in one-to-one negotiation or auctioned as an individual or as part of a matched team. It is hard to imagine any other commodity being sold one unit at a time in anything approaching the hundreds of thousands of horses sold annually. Biological individuality created individual animals, a fact that imposed some limits on commodification. Choosing individual horses was a fine art, requiring years of experience in a marketplace where caveat emptor dominated. Breed, color, and size were easy to see but might not fully relate to the employability of a horse. Even those qualities might be deceptive. Horse dealers could and did dye hair and overfeed animals just before sale, since a shiny coat and "firm flesh" supposedly indicated health. Another ploy involved feeding pepper to an old horse to make him look younger and livelier. Sellers were very good at disguising lameness and sore legs, often by resting a horse for several weeks before sale. All stables contained a ring where horses had to demonstrate their gait, but even that could be fudged by covering the ring floor with soft bark.[2]

Even with full disclosure and no fraud, buying horses was an inexact science. The devil lay in the details. Pounding urban pavements was hard, and a horse with strong legs on the farm might still be prone to lameness in the city. Some horses were more susceptible to respiratory infections but perhaps only in the winter, not at the time of sale. A horse buyer might listen to the lungs, hoping to detect an irregularity, but that was imprecise. The genealogy of a horse might tip buyers off to potential problems but was rarely known with any assurance. Strength varied not just from breed to breed but from horse to horse and load to load. One writer reported a brewery horse, chronically lame when pulling four tons, who gave excellent service when pulling only three and a half tons.[3]

Judging the "character" of horses was very difficult at an auction, but stables wanted to avoid "vices" in equine employees (and, for that matter, human employees), such as laziness, timidity, or viciousness. Some emigrant horses could not stand the noise and crowding of big-city life, but no buyer could judge that. Some horses were never "smart" enough to figure out how to back up over a sidewalk grating to unload coal or beer. Here, once again, horses were more than just machines, since "character" varied from individual to individual. Horses, like humans, could be fickle, and the marketplace recognized this.[4]

Age was a crucial market variable, since a five-year-old might have twice the productive life of an eight-year-old. Stable manuals were full of dental hints to determine age based on the size, shape, and even pitch of the teeth, but age remained pretty indeterminate, especially since sellers sometimes filed or painted teeth. Some sales techniques were merely cosmetic, such as braiding tails or putting ribbons in the mane. The Chicago Stockyards actually employed nine barbers to groom horses for sale.[5]

Buyer preferences varied also. Teaming companies required horses matched in size and strength. In mismatched pairs, one invariably worked harder or responded faster than the other, often leading to harness sores, besides being inefficient for pulling purposes; thus, sellers marketed most draft animals in matched teams. For horses, as for humans, some couples got on better than others. Style also played a role. Even freight companies preferred teams with matching colors, assuming that color conveyed compatibility. Undertakers wanted all-black horses and breweries Clydesdales, essentially for cosmetic reasons. One New York furniture mover preferred slow horses because they were less likely to run away and smash his load. There was even a market for blind horses, who could work in mines (mine horses rapidly went blind anyhow) or turn mills. Tastes could change rapidly. Elite carriage owners wanted imported Cleveland Bays in the 1880s. When Mrs. William Vanderbilt, the famous New York socialite, imported two Hackneys in 1891, they became the rage, and Cleveland Bays could be sold only to cab drivers at a much lower price. For most of the century, goods haulers preferred Percherons as draft animals, but Belgians became more popular in the 1890s.[6]

The tales of chicanery in the horse market are often exaggerated, and fraud happened most often when individuals sold used animals in informal street sales.[7] At the bottom of the market, every large city had a stable with a poor reputation, like the Skinner's Market in New York City, where "at halter" sales took place. This market was notorious for selling diseased horses and took its name from hide dealers who bought otherwise unsellable animals for one to three dollars, planning to kill them for their skins. It had a reputation for "cheapness," as is reflected in the comment that "the horses are low and the buyers and sellers rank in the same category" (*New York Times,* December 26, 1869).

While hard to enforce, buyers in such markets had common law protection against such fraudulent techniques as the use of false bidders to puff up the price and dyed hair. There were also different levels of sales stables with various levels of warrantees. Tattersall's, the most elite stable in New York, offered its buyers especially strong protection. All cities had a large wholesale stable with rules like those in the Chicago Stockyards. Sellers had to warrant the health of their animals, although with limits described in heavily coded language. Horses could be sold as "sound" (guaranteed not to be lame), as "workers" (wind guaranteed), or "at halter" (meaning "as is"). Describing a horse as "a little bluish in one eye" voided any warrantee that it could see out of both eyes. Almost always, sellers guaranteed horses as "free from vice," meaning that they were tame and usable. This could lead to considerable argument, since a horse might be tame for one driver but not another. Even "at halter" sales carried some implied warrantees, and purchases could be returned if concealed defects were discovered within twenty-four hours.[8]

Wherever they bought, large-scale or elite purchasers usually insisted on the right to return animals within a specified time period, usually twenty-four or forty-eight hours. Buyers could engage in a practice called "bushing," demanding a rebate from the seller of a newly purchased horse in which some supposedly unsound condition had been found, rather than returning it. Apparently those who practiced "bushing" thought that they could still get some value from the horse, if only by reselling it. Police and fire departments paid handsomely for thirty-day "no questions asked" return privileges. There was no way for them to tell if horses were suited to their special needs without a long training period.[9]

Arguments could be settled in a number of ways, most obviously through the courts, but legal action could be slow and chancy, especially if an out-of-town seller had returned home. Commonly, the large sales stables insisted that sellers agree to a binding arbitration process. This encouraged buyers to patronize the commission dealers who sold at the large auction houses. Advice manuals told buyers to have purchases checked by veterinarians, and so the new profession was dragged into this messy process. Veterinarians were especially valued after they could detect animals with glanders, a deadly disease, which often went into remission. Dumping glandered horses was a common ploy of unscrupulous sellers. Henry Bergh, the president of New York's ASPCA (American Society for the Prevention of Cruelty to Animals), once caught a salesman blocking the nose of a glandered horse (a runny nose was a symptom).[10]

Informal local markets seem to have been the norm until the 1850s. Railroads allowed long-distance shipment of horses, sixteen to twenty-four to a car, and there were regional variations in production that made interregional trade profitable. By 1893 horses cost $93.37 on average in Rhode Island, the most urbanized state in the nation. In comparison, they were worth less than $25 in the South-west and $43.40 in Illinois, the leading breeder of heavy horses. Such comparative advantage guaranteed a national market. There was always some buying directly from the farm, especially by large firms that could employ and send out professional purchasing agents. For example, a New York firm that needed to replace more than a thousand horses lost in a fire dispatched buyers "directly to the west."[11]

The normal process involved a middleman buying from a farmer who had fattened the animals and begun training them, then sending a carload to Chicago for sale to a commission merchant, who would sell the horses to the agent of a sales stable in another large city, where they would be sold at a retail market. Every step in the process, except the first (haggling at the farm or railhead) and last (street sales), was through auction. Auction houses typically got 10 percent of the price. Prices varied not only from animal to animal but also from year to year. Rapid price fluctuations made buying more complex. Consider this price series for urban horses: 1880, $54.75; 1889, $72.00; 1894, $48.00; 1895, $31.00; 1900, $49.00. Seasonal variations were also important, with prices higher in the spring than the fall or winter.[12]

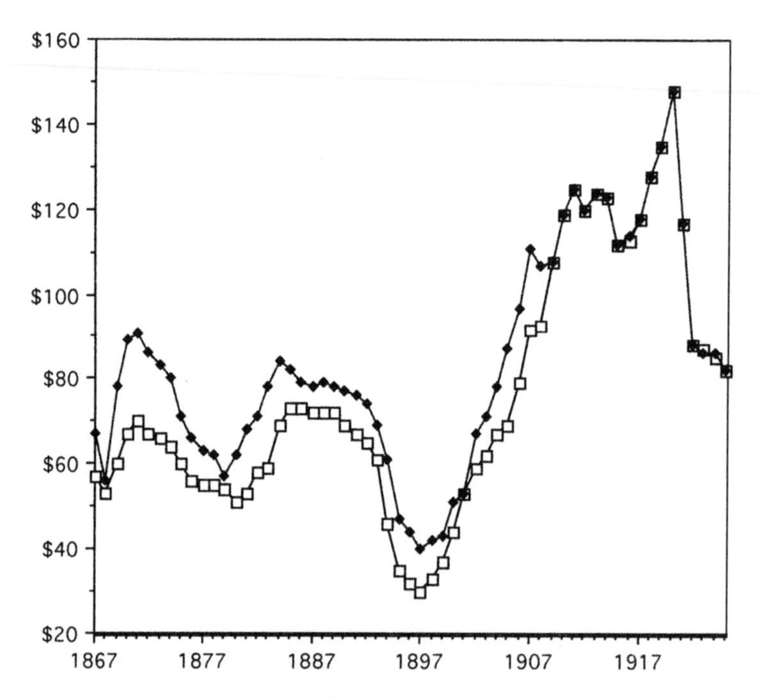

The Value of Horses and Mules from 1867 to 1925. *Open squares,* horse values; *solid diamonds,* mule values.
Source: Historical Statistics of the U.S. (Washington: GPO, 1975), 520–21.

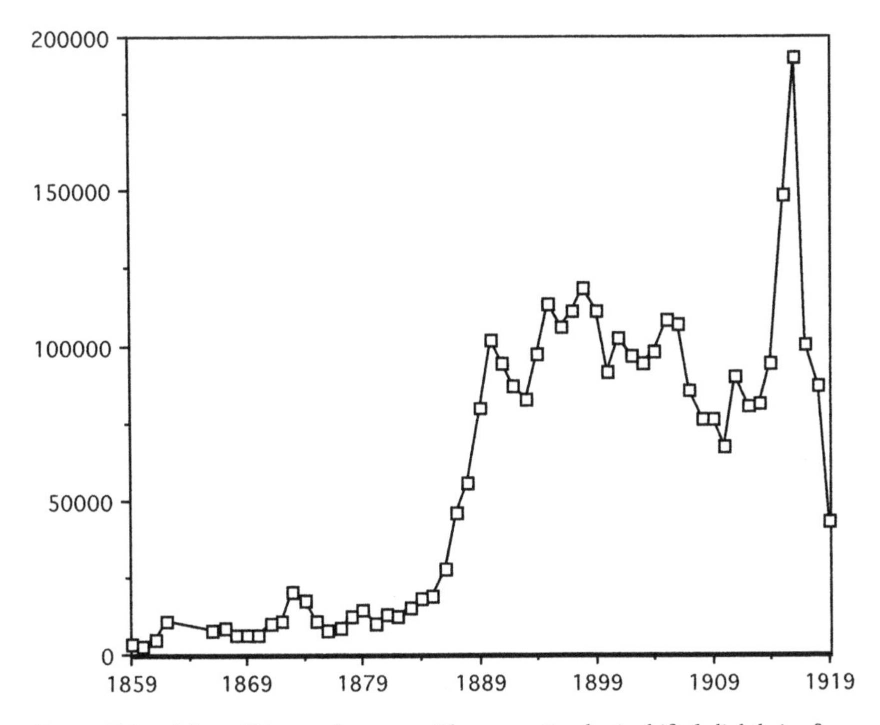

Horses Shipped from Chicago, 1859–1919. The accounting basis shifted slightly in 1879. *Source: Annual Reports of the Chicago Board of Trade.*

The system ultimately centered on the stockyards at Chicago, the largest horse market in the world, where up to 180,000 horses a year changed hands. John Flinn's nineteenth-century paean to Chicago, *Chicago: The Marvelous City of the West* (1891), stated that the Union Stockyards were the city's largest, with four thousand stalls. Auctioneers sold sixty horses an hour daily, with some horses changing hands more than once.[13]

Horse prices declined in the 1890s, an event that experts attributed to the electrification of street railways and a bicycle fad but that probably owed more to the general deflation of the 1890s and to overproduction. The price decline was so bad that some young horses were killed for the value of their hides and other by-products.[14]

The horse industry then turned to foreign markets, dramatically increasing exports to Europe and South America. By 1897 the French and Prussian armies were each purchasing ten thousand horses a year in the United States. An 1898 congressional investigation looked at the dimensions of the international market. The number of foreign purchasing agents at the Chicago markets had increased

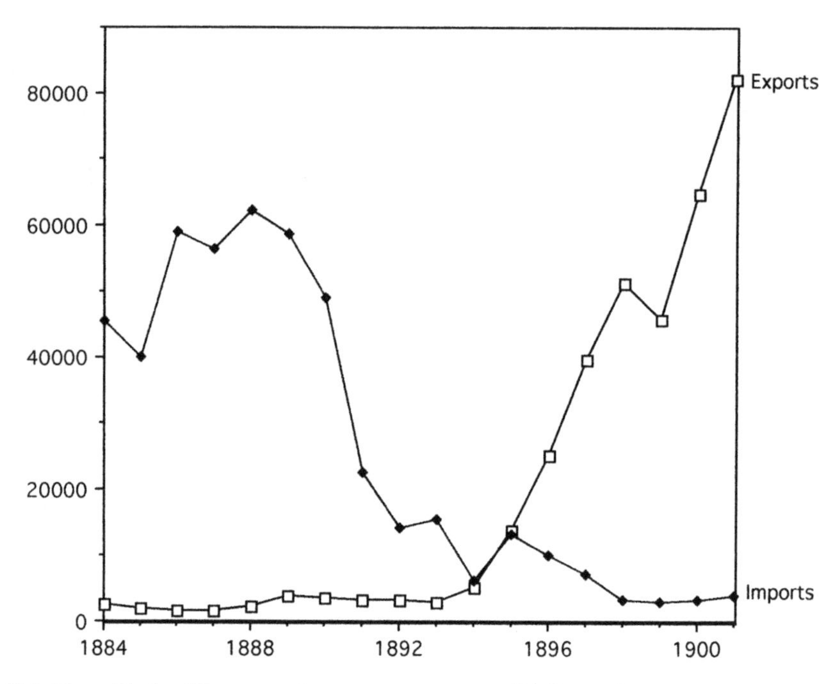

U.S. Horse Trade, 1884–1901. *Open squares,* exports; *solid diamonds,* imports.

from two to seventy since the crash five years previously. Britons bought twenty thousand American horses. The London General Omnibus Company bought many horses discarded by rapidly electrifying U.S. trolley companies. Even American breeding stock was going overseas. American horses had a price advantage, even after a \$30–\$40 transportation charge, to the point that European producers sought tariff protection. When the British military was purchasing eight thousand American horses a month during the Boer War, it sent a team of officers scurrying through the American countryside. John Flinn noted that horsemeat canneries springing up in the United States were geared toward exporting to Europe.[15] This was a fast and flexible response to a changing market, with an important shift coming in 1896, when the United States, historically an importer of horses, exported twice as many horses as it took in.

Local Markets

New York provides an outstanding example of local markets in large cities. Several five-story stables near Third Avenue and Twenty-fourth Street sold living machines fresh from Chicago to dealers from smaller cities, to large stables, and even to a few individuals. By 1886 these large, local wholesale stables were selling to retailers more than 100,000 horses valued in excess of $15 million a year. The actual venue for many transactions was a saloon called the Bull's Head. The wholesalers sent leftover horses to a public auction house. Many sales took place at other locations, sometimes through the agency of a livery stable owner, who might even take a horse on consignment. Considerable one-to-one trading also occurred, often at informal street sales usually held at stated places and times. Street railways often had an annual sale of horses that they had fully depreciated—those who had worked for more than five years, since experience showed that horses were much more likely to become sick after that time. Every fall from 1879 to 1894, the Providence street railway, which depreciated its living machines faster than did most firms, auctioned all horses that had worked for three years. Usually freight haulers had their own depreciation rules.[16]

The urban horse commonly had a career that began with a training period doing light farm work, when the horse was broken to harness; then he hauled for a street railway firm, which sold him to a delivery firm, which sold him to a cab firm, which sold him to a peddler, in much the fashion of the fictional Black Beauty. Elite carriage owners also sold their animals after they became old or unfashionable. Peddlers and cabbies generally had the oldest, least sound horses. Some horses that became chronically lame on pavements were resold to farms, as were other horses that were poor feeders but that might fare better in a grazing environment. Lameness and then the veterinarian's pistol was the most common end for city horses.

Commodifying Horse Wastes

Even the wastes of horses were commodified. The collection of urban manure had old, even ancient roots. Again, the process is most easily documented in New York City. Before 1787, individuals roamed the streets and picked up manure. In that year the Common Council supposedly sold an exclusive license to a William Hitchcock, who sold the street sweepings to farmers for fertilizer. Street sweepings varied in quality and were worth more if from an asphalt street than if from a gravel street or a dirt alley. They were always worth less than stable manure, a purer product. The older pattern of individuals collecting street manure for urban gardens never fully went away, and as late as the first half of the twentieth century neighborhood children in the Italian American neighborhood of East Harlem did a thriving business collecting horse manure from the streets for backyard gardens in the area.[17]

Heavy urban animals produced between thirty and fifty pounds of manure a day, probably averaging around seven tons a year. Selling this manure to farmers at appropriate seasons of the year was a major contribution to the balance sheets of stable owners. In New York licensed carters originally held exclusive privileges for manure carrying, but after the licensing system collapsed in the 1830s intermediaries paid stables for manure to resell to farmers. The farmers then hauled the product home themselves, often on the return trip from delivering their produce to market.[18]

High-value stable manure could significantly enhance the income earned by an urban horse owner. Just in their role as manure producers, many urban horses earned their purchase price back in five or six years. The owner of the street railway in Louisville claimed at a convention that he had trained his horses to leave their droppings in the stable yard, instead of city streets. His peers ridiculed him, but the claim itself shows the value placed on manure. Other managers did train animals to visit the manure pile before and after working shifts but evidently did not claim complete excretion training.[19]

This recycling, perhaps better viewed as an exchange of energy between country and city, did not come without drawbacks. Urban residents complained to boards of health with increasing frequency in the late nineteenth century about manure piles. Urban populations, both horse and human, grew in size and density, making bad smells more common and more noticeable. In the 1850s, for example, nobody in Brooklyn protested about manure carts, but in the early 1880s, as Brooklyn became more suburbanized, the residents of the main streets began to complain about them. New York City required closed manure carts by 1871 and Brooklyn in 1882. The same fears of bad smells that had contributed to the abandonment of privies for more sanitary or at least better smelling water closets, the contents of which cities had once sold to farmers, applied to stables. City residents did not want their neighborhoods smelling like a barnyard.[20]

Manure disposal seems a classic urban NIMBY (not in my backyard) issue. Nineteenth-century ideas of health required the immediate removal of manure before bad smells permeated a neighborhood. Stables had to remove the manure for the health of both horses and humans, but piling it up someplace for transshipment just generated the same complaints in a different location. Although draining stables into sewers probably reduced the problems associated with urine, there was no equine equivalent of the technologies (water closets, sewers, and sewerage treatment) put in place for humans as their wastes lost value.[21]

A decline in the price of manure worsened collection, and complaints grew louder late in the century. The Second Avenue Street Railway in New York City reported a gradual decline in the annual value of manure earnings from $4.60 per horse in 1860 to $1.10 in 1885. As early as 1884 one New York City stable keeper complained that manure, which he had sold for fifty cents a load the previous year, now had a negative price—that is, he was paying a dollar a load to have it removed. Of course, part of the decline was the long-term deflation of the late nineteenth century, but more of it reflected changes in the fertilizer trade. Guano (decomposed bird excrement) had become a valuable, low-priced fertilizer in Europe in the 1850s but had little effect in the United States, since Americans had access only to inferior supplies. In the 1860s, American guanos, enriched with fish scraps, became available, but even these could not compete with urban manures in urban hinterlands. In 1867 the Pacific Islands Guano Company (Baltimore) acquired rock phosphate (geologically ancient guano) deposits in South Carolina. Industrial chemists continually improved the quality and reduced the price of rock phosphates, ultimately destroying the manure market.[22]

Commodifying the Dead Horse

Even death became mechanized in the nineteenth century. In death the horse became a commodity as well. When horses became sick, lame, or just too old to justify the money spent in feeding and stabling them, their owners had them shot. It is likely that relatively few animals died directly from natural causes, although some did drop dead on city streets or in the stable. Owners insured horses, a major capital investment, but insurance companies were reluctant to let owners shoot their own horses, since the temptation to make an insurance claim when business was slow might prove too much for some owners. Veterinarians, police officers, and agents of anti-cruelty groups could shoot horses, however, if they deemed the animals diseased, dangerous, or incapable of further work. This would validate an insurance claim. Between 1887 and 1897, New York's ASPCA, the city's leading killer of animals, shot between eighteen hundred and seven thousand horses a year. (This is the beginning of the familiar paradox by which humane societies have become the leading killers of animals in cities.) Vets accepted animal owners' valuation of animals almost exclusively from their productive utility.[23]

As usual, New York had the most massive problem. Seven to eight thousand horses a year died in Manhattan in the early 1880s, compared with only fifteen hundred a year in Chicago. As many as thirty-six died on Manhattan streets daily during the Great Epizootic in 1872. One writer has estimated that twenty-five hundred horses died in New York during the epizootic, although this figure is much higher than anything that we can glean from contemporary reports. In 1910, probably the peak year, the municipality reported the removal from its streets of more than 20,595 dead horses.[24]

Marketplace considerations dictated prompt removal, since decomposition could ruin many by-products. Stable owners had contracts with rendering plants for the removal of dead animals. City governments also had contracts with rendering firms to remove dead horses from the streets. In New York City, for example, the New York Rendering Company, which had a monopoly on public and private rendering, guaranteed to the Board of Health removal within three hours. The municipality was not alone in wanting rapid removal. After rigor mortis set in, hides lost their value, so speedy pickup was also important to the company, which made the highly improbable claim that it removed most animals in fifteen minutes. Health or police officers notified the firm of dead animals via the city's new police telegraph system. We have not seen any complaints about slow removal in *Health Department Reports* or in New York newspapers, probably because the interests of the rendering firm and of the Health Department coincided. The picture of a dead horse in a gutter that frequently appears in American history textbooks shows an anomaly. Carcasses traveled on special wagons with ramps that tilted down to the street. A horse-powered winch pulled a cable that hauled the dead animal up onto the wagon's bed. A tarpaulin covered the dead horse en route. Other municipalities had removal systems like that of New York.[25]

This process differed widely from that in those countries of continental Europe where consuming horsemeat (hippophagy) was commonplace. In Paris, owners seldom killed horses, since the authorities forbade processing dead or diseased horses into human food. Rather the horses were driven to the slaughterhouse, some limping along on three legs. Owners usually shaved the horses before getting rid of them, preferring to sell the hair themselves rather than letting the slaughterhouses benefit from this valuable by-product. The Parisian public found these pathetic parades of bald horses very offensive.[26]

Most Americans followed the British folkway of not consuming horsemeat, so they did not process horses in the Parisian manner. There were some American intellectuals who advocated hippophagy, notably Henry Bergh, founder of the American Society for the Prevention of Cruelty to Animals. Bergh believed that death in a slaughterhouse was preferable to death under some peddler's whip. This was a most exotic position—he advocated the consumption of horsemeat not for the benefit of human populations but for the horses themselves, a kind of euthanasia. The first dean of the University of Pennsylvania Veterinary School served the unidentified remains of Dora, his pet horse, to his colleagues at a formal dinner just to show that the opposition to hippophagy was purely cultural. The prominent veterinarian John S. Billings argued that horsemeat was healthy, that aversion to it was based on "ignorant customs," and that many urban horses were more suited "for food than work."[27]

Some American humans did eat horsemeat, mostly in city neighborhoods populated by immigrants who had eaten it in the old country. Clay McShane can remember "pork shops" in German neighborhoods in New York City in the 1950s, which sold horsemeat. Evidently, consumption of the meat increased during World War II rationing. There were earlier examples. During the siege of Vicksburg, mule steaks commanded twenty dollars apiece, and other soldiers consumed horsemeat at various times during the Civil War.[28]

The variety of products made from dead animals was remarkable. Rendering plants shaved the hair to be used for cushions, "thus the dead are made to minister to the comfort of the living." Hair also became a stiffener for plaster and was made into blankets. Skinners cut the hide off, using the rump portion of the hide for highly valued cordovan leather. They boiled hooves to extract oil, especially for glue but also for gelatin. Renderers boiled the carcass in a pressure boiler to separate flesh from bones and then carved the leg bones into knife handles and combs. The ribs and head were treated to remove oils and then burned for bootblack, a substance valued not just for polish but also for filters in sugar refineries. Vapors from this process became the chief source of carbonate of ammonia and a valuable insecticide. The phosphate of lime extracted from bones was processed into phosphorus for matches. Horsemeat became pet food. The remaining mass was pounded with potash to produce prussiate of potash, which was needed for dyes and poisons. Fats skimmed off the top of the vat became soap or candles. One St. Louis rendering firm in 1896 claimed a twenty-four-dollar profit on each carcass.[29]

Before the beginnings of modern processing, all but the hide and fat was cut up, perhaps baked, and then hauled to the country to be dumped on fields as a fertilizer. By the 1820s it was common to apply this meat, and especially dried blood and ground bones, as fertilizers. The pioneering 1840 research of the German chemist Justus von Liebig demonstrated the value of this folk custom, especially for the latter two items. In 1851 D.J. Browne's *American Muck Book* suggested that the nitrogen-rich remnants of soap making should be applied at a rate of two hundred pounds to the acre. In spite of the possibilities of recycling, however, as late as 1880, Charleston, South Carolina, was still dumping dead animals on the ground outside the city, while Albany, New York, was throwing them into the Hudson River.[30]

The new rendering factories associated with large slaughterhouses starting in the 1850s processed carcasses from not only horses but also hogs and steers. Dr. R. W. L. Rasin, the Baltimore chemist of one of the phosphate fertilizer firms, began experimenting with mixing the rendered slaughterhouse wastes (or, more properly, the "soup" left after the fat had been skimmed) with rock phosphate, thereby increasing the nitrogen content of the fertilizer. By 1868 he was buying tanks of slaughterhouse wastes from the huge Chicago meat packer Armour & Company, wastes previously dumped into sewers. In 1871 a New York rendering firm sold eighty-three tons of soupy wastes for $38.21 a ton. Only the giant new slaughterhouses or, in Boston's case, abattoir (i.e., public slaughterhouses on the European model) had the economies of scale to fully recycle dead animals.[31]

Rendering was a process that reeked horribly. It raised all of the NIMBY problems that manure pits did. The *New York Times* described the air around one plant as "poisonous." Even in the colonial period, blood boiling, candle making, and soap manufacturing had been regulated as noxious trades, usually by forcing them to the outskirts of town. This was much harder to do with the huge meat-processing plants that appeared in the 1860s. Firms processed wastes in giant boilers, sometimes outdoors, since they did not like the cost of closed boilers (autoclaves) and feared explosions. A local nuisance could become citywide. On July 21, 1873, the John P. Squire Company slaughterhouse in Cambridge, Massachusetts, created a very large stink from rendering late at night, causing residents from Radcliffe College to Boston's South End (a distance of five miles) to wake up nauseous and to vomit. Ironically, conditions were worse that night because Squire's had made a mistake while trying newly required pollution control equipment. Some relief came with the imposition of more effective health regulations in the 1870s, most notably requirements that rendering take place indoors in autoclaves, not vats. The development of air pollution control devices, which passed exhaust through superheated steam consuming much of the residue, reduced a sometimes-citywide stink to a neighborhood aroma. The areas around such plants typically became neighborhoods for poor, working-class residents probably willing to trade good air for slaughterhouse jobs.[32]

The Horse as Consumer

Just as a vast industry of manufacturers, showrooms, repair shops, garages, petroleum refineries, and gasoline stations has grown up to serve the needs of the automobile, a support infrastructure existed for the living machine. Directly or indirectly, the urban horse exercised a powerful multiplier effect on both local and national economies. A horse-based economy demanded enormous inputs of land, labor, and capital. Humans equipped with expensive machinery had to construct work tools for horses (such as harness and vehicles), stables, and roads. The horse-powered economy required massive amounts of land, both for grazing in rural areas before migration and for food supplies after. Stables occupied expensive urban land. Cities had to change street plans to provide more space for vehicles. Stables required "maintenance" employees, most obviously hostlers (grooms), but also such specialized occupations as farriers, horseshoers, and veterinarians.

The quantity of workers directly involved with the living machines was also huge. The number of teamsters in the United States increased from 120,756 in 1870 to 368,000 in 1890. Teamsters were not the only people who earned their living from the horse economy. Urban stables usually employed one hostler (groom) for every sixteen horses. The number of hostlers in the United States increased from 17,586 in 1870 to 54,036 in 1890. Very large stables had other employees as well: carriage painters, horseshoers and wheelwrights, purchasing officers, accountants, managers, and so on. There were 8,504 Americans who identified themselves as livery stable keepers in 1870 and 26,737 in 1890. The 5,103 street railway employees in 1870 had increased to 37,434 in 1890. Both the manure and the rendering trades also employed humans and capital.[33]

On a national basis the number of support industries and establishments needed to maintain the horse-powered society was extensive and varied. In terms of employment and capital invested, the carriage and wagon industry was most important. This industry was concentrated in the Middle West, New England, and the Middle Atlantic states, with Ohio, Indiana, and New York being the leading states in terms of value of product and number of employees. While in 1905 only 38.6 percent of the total number of establishments manufacturing carriages and wagons in the United States were located in cities with a population of more than twenty thousand, they employed more than half the workers and generated more than half the value of product in the industry.[34] The horse-drawn society also required manufacturers of blankets and other "horse clothing," makers of saddles, harnesses, and whips, and horseshoers and wheelwrights. As the society became more dependent on the horse, the amount of capital invested and the number of workers employed in these industries vastly increased. In 1890, 9,163 establishments manufactured carriages and wagons or their parts, employing more than 90,000 workers to make over one million vehicles worth over $32 million. As late as 1909, 33 establishments employing 1,830 workers manufactured horse clothing, while 57 firms employing 1,946 workers manufactured whips. Nineteen firms specialized in horseshoes.[35]

The horseshoe trade reflects much of the innovation and specialization in support industries. In 1817 Brooklyn resident E. Maynard took out a patent on horseshoe calks, the equivalent of a cleat on a modern athletic shoe. Calks like these were one of the keys to adapting horses to cities, allowing hooves to grasp the cracks between paving blocks, giving extra leverage, especially on hills. Maynard may have filed prematurely, since there were very few block pavements for heavy hauling before the 1850s. Nobody filed additional patents until the 1860s, when a dozen more, including several for detachable calks, were filed, as were thirty-two patents for machines to make or sharpen them. Patent office records also show hundreds of patents for horseshoes, nails, and the equipment to manufacture and fit them before 1874 but few before 1855. The making of horseshoes, which had for many years been the province of neighborhood blacksmiths, became an activity of large iron foundries. In cities, specialized horseshoers replaced blacksmiths for fitting and replacing shoes, except in small stables. In 1887 a major New York stable employed 18 full-time horseshoers, one for every sixteen horses. In Boston 238 horseshoers pursued their craft in 1900.[36]

Cities had many businesses supporting the needs of horses. In 1855, for instance, the town of Davenport, Iowa, which had a population of about eleven thousand, had 76 blacksmiths, 46 carriage and wagon makers, 2 carriage trimmers, 28 draymen, 21 livery men, 20 saddle and harness makers, and 145 teamsters employed in servicing the horse trade. In 1870, Boston, with a human population of approximately 250,000, had a very elaborate horse service industry, with 62 carriage dealers, 61 firms making harnesses, 29 hay dealers, and 15 wheelwrights. By 1900, with the human population of the city at 561,000, these numbers had increased to 105 carriage dealers, 99 harness makers, 51 hay dealers, and 30 wheelwrights. The *Boston City Directory,* not a very exact source, listed 60 livery, boarding, and sales stables in 1850, 106 in 1870, 190 in 1890, and 192 in 1900. Pittsburgh, with a population of over 238,000 in 1892, had 63 sellers of flour, feed, and grain, 56 harness makers, 81 horse shoers, and 17 carriage manufacturers.[37]

Moseman's Illustrated Guide for Purchasers of Horse Furnishing Goods, Novelties and Stable Appointments (1892) illustrates the extent of the trade in horse goods, both retail and wholesale. C. M Moseman & Brother, a firm that evolved out of a feed concern into a general supplier of horse goods, published this catalogue. After a modest beginning in the late 1860s, by the early 1890s the firm occupied a five-story building at 128 Chambers Street, just west of Broadway, where most of New York City's leading saddlery and harness wholesalers were concentrated. The building had several showrooms for horse goods, including a 110-foot basement room that displayed 275 different styles of harnesses.[38]

Moseman & Brother specialized in the finer quality of horse goods, as reflected in their boasts about the store they established in 1901 at 571 Fifth Avenue: "There is not a finer nor more swell establishment of the kind in this city or any other." Eventually the firm opened branch offices in Berlin, Paris, Moscow, Vienna, and Walsall (near London), England. The catalogue went through five editions and sold for one dollar. Goods could be bought at the store or ordered through the catalogue. Among its categories were "stable requisites," such as pitchforks and shovels, brooms, many versions of horse brushes, curry cards, combs, and "tail squarers." Horse clothing such as boots, caps, and tail covers came in many designs and colors, as did sands for floor monograms and crests. Buyers could purchase harnesses of many varieties, not only for single carriage horses but also for trotting racers, for tandems, for heavy draft horses, and for mules. More than two hundred types of riding and driving bits were available, over a hundred different types of whips, and over fifty types of saddles. Choices for the carriage included a variety of jacks, lamps, odometers, wheel guards, and driving sundries and for the stable a large variety of clippers and grooming machines, as well as oils, dressings, and veterinary materials.

The horse-related products, as reflected in the variety of goods offered by Mosemans, often had large lines of goods associated with them. For instance, Rogers Peet & Company specialized in livery of various kinds. In their 1904 catalogue, they featured eleven pages of coats for coachmen and ten pages of coats for grooms. These included breeches, boots, cuffs, gloves, scarves, trousers, waistcoats, and crepe bands for "dress livery" and similar goods in "undress livery" for "real" country use. In their "prefatory note," the firm noted that "men choose livery as they choose horses—for style, endurance and other good points."[39]

Another commodity consumed by the horse was land, and a large portion of land in the United States, both rural and urban, served horses' needs. For one thing, horses ate a lot, a subject we will take up in more detail in chapter 6.[40] At this point we will note only that the 1900 urban horse population of approximately three million consumed about 7,700,000 tons of hay and 8,800,000 tons of oats. To grow this amount of fodder may have required as many as 12.2 million acres, roughly four per horse.[41]

Numbers of firms and workers were required to grow the crops, move them to cities, and distribute them at both the wholesale and retail levels. It is impossible to quantify the number of people or the amount of capital tied up in these equine needs. In 1900, 4.5 million American farms raised horses as their primary product, with roughly 10 percent of those animals destined for urban customers.[42] At any time there were probably as many horses on the farm being prepared for urban hauling as there actually were in the city, since most animals moved at the age of five and the life expectancy of city horses was roughly until age ten. Most mares in the major draft breeds spent their lives on the farm, invariably pregnant. In 1895, the peak year, 113,193 horses were shipped from Chicago. For each of those horses, there was a mare on a farm somewhere.

The living machine required city land, and not just for stables. All large cities designed street systems suitable for horses. Sidewalks, the norm after 1850, separated vehicles from pedestrians. At the end of the Civil War, General Ulysses Grant claimed that his only political ambition was to return to his hometown of Galena, Illinois, to put in sidewalks, a sign of the emergence of horse traffic. Gridiron plans, popular in the mid-nineteenth century, eliminated turns too sharp for a horse with a four-wheel vehicle. For example, Boston's Beacon Hill neighborhood, an upper-class subdivision first planned in 1795, when people walked or perhaps drove two-wheeled gigs, had streets 30 or 40 feet wide from building line to building line. Its most prominent internal artery, Mt. Vernon Street, was 57 feet wide. Its most important shopping area was on Charles Street, 60 feet wide. By contrast, the Back Bay, another upper-class neighborhood, but one built up during the age of the four-wheeled landau, had streets 60 feet wide (not to mention 16-foot-wide alleys). Commonwealth Avenue, its most imposing residential street, was a whopping 200 feet across, including a park in the middle. Commercial Boylston Street was 80 feet wide.[43] Other public improvements of the mid- and late nineteenth century also suited horse traffic. Drives in the new ornamental parks allowed the display and sometimes racing of carriages, whose owners were among the most important lobbyists for such improvements.

Pavements also served the needs of horses. Cities adopted macadam gravel pavements on lightly traveled roads in the 1850s and added stone block pavements, which were easily grasped by calks, on heavily paved streets. At least two other pavement types were rejected at the same time because they did not suit horse transportation. Concrete shattered under the constant blows of horse shoes—its adoption awaited the arrival of rubber-tired motor cars—and wood blocks posed a peculiar problem. They absorbed urine and bled it out as ammonia on hot days. Of course, both of these pavements had other undesirable characteristics—cost for the former and durability for the latter. Asphalt, the preferred pavement by the 1890s, offered an important characteristic—smooth pavements could be swept by rotating brooms pulled by horses, making it cheaper to clean up the mess left by other horses.[44]

As one thinks about the horse not as an animal but rather as a living machine in an urbanizing society, its role in the process of commodification becomes clearer. Horses had value assigned to them from their very birth. In the nineteenth-century city this value related primarily to their usefulness for work—as prime forces driving horse whims and treadmills, providing power for lifting, and as haulers of drays, wagons, omnibuses, and streetcars. Horses themselves were major consumers of foodstuffs necessary to provide fuel to supply them with energy and of all kinds of harnesses, clothing, shoes, and other kinds of furnishings. Because so much of the urban built environment was shaped by their needs—for example, stables for housing and streets shaped to facilitate movement—they can even be thought of as consumers of these features of the city. Even their manure was of value as fertilizer, while at death their hides and hair were transformed into useful products. Thus, horses, as living machines, assumed a critical role in the emerging capitalistic economy of nineteenth-century urban America.

CHAPTER TWO

Regulation

Controlling Horses and Their Humans

The 1850s were a decade of rapid urbanization in the United States, and city growth demanded living machines. Although exact data on the total number of urban horses is unavailable, the one city for which reasonably reliable data does exist—Boston—shows nearly a doubling of its horse population. In addition, the number of urban teamsters in the nation more than doubled, as did the value of carriages produced, clear indications of increased horse usage.[1]

Ownership and operation of the horse became more centralized, most obviously as street railways replaced walking but also as express companies and other delivery services replaced individual carters. This centralization implied overarching organizational structures and new behaviors by all concerned to control the new applications. An urban hackman probably stabled his horse in a backyard shed, but a street railway had a stable for hundreds, if not thousands, of horses. The same difference existed for carters and express companies. Even in the leisure use of horses, expanding urban elites increased the number of privately owned trotters. They were more likely to be stabled in centralized boarding stables. So most horses, once managed by individuals, were now often controlled by bureaucracies, and increasingly ownership and driving functions became separated.[2]

Freight Movement

The control of living machines occurred at different levels: owners, managers, teamsters, private anti-cruelty groups, and municipal agencies all had roles in the complex, sometimes contradictory, process. Day-to-day, minute-to-minute control of horses rested with drivers, who were sometimes their owners but were more commonly employees. Obviously, owners sought to control their drivers. Drivers were subject to public regulation, often enforced through quasi-public groups like anti-cruelty organizations and licensed veterinarians. Finally, there were peer groups, perhaps organized into unions, which also sought to control driver behavior.

Before the beginnings of the horse revolution of the 1850s, freight in cities had largely relied upon two-wheeled carts whose owners, individuals licensed by city governments, walked alongside them. Carts had technical limitations. They were slow, especially because cities usually required that they be led by hand, and they held fewer goods than wagons, even when only one horse pulled a wagon. Carters would sometimes stand in the back when driving the vehicles, a dangerous arrangement, since two-wheeled vehicles tended to be unstable, as accounts of accidents attest. Carts dominated urban traffic until the 1850s, and they were probably still an important element of freight traffic as late as 1875. One horse pulled most carts, carrying light loads like dung or furniture. Carts offered the advantage of easy loading and unloading, since they could be tilted backward to empty cargo. They endured well into the twentieth century for loads requiring dumping, like coal, garbage, or soil from construction sites. A few were adapted for special purposes. A cart mounted with a tank peddled supposedly pure water for tea in New York in the 1790s, but typically carts were general-purpose, not specialized, vehicles.[3]

Most of what we know about carts comes from New York City sources, but there is no reason to believe the pattern was much different elsewhere. The New York City Common Council licensed carts (sleds in winter), limiting size and rates. Cartmen (we have not found a single cartwoman) could own only one horse and one cart. They had to know local geography, farriery (a premodern craft combining blacksmithing and folk veterinary medicine), and driving. The carters operated not unlike a medieval guild and wore white frocks as a quasi uniform or emblem of the trade. May 1 was Carter's Day, when they marched through streets with horses bedecked in flowers. The number of carters expanded greatly with the growth of the city: there were 321 carters in 1785, more than 3,400 in 1835, and 8,500 in 1855. Hackmen also expanded in numbers, although by a smaller amount. Carts increasingly displaced wheelbarrows, which had been widely used in retail businesses because retail customers spread themselves over larger areas. The city had banned blacks, "youths," and immigrants from the trade after it began licensing carters in 1729, allowing those groups to carry only "dirt" (i.e., human and equine manures). The restrictive characteristics of the system, however, broke down under the onslaught of Jacksonian egalitarianism, and in 1828 the Common Council increased the number of licenses to allow the formerly barred groups to enter the trade. Previously all carters had been independent entrepreneurs, but some now began working for companies. The licensing system gradually disappeared under pressure from the new express companies of the 1840s, since it increased their operating costs. Similar patterns can be observed in other cities. Haulers became increasingly Irish, and so-called independents, who owned their own horses and carts, became less common.[4]

One carter, Isaac Lyon, published his reminiscences. He described his peers as craftsmen who earned their pay with their knowledge of the city and careful attention to loads entrusted to them (he specialized in "fine arts," carrying objects like paintings and mirrors). According to Lyon, "A Cartman should be an encyclopedia and an intelligence office together." To him, practitioners were idealized republican workmen, noted for their patriotism, always turning out in their white smocks and highly polished black boots for parades, and serving as volunteer firemen. These practices extended beyond New York. In Pittsburgh, carters in white smocks standing on their vehicles highlighted labor parades. Lyon bragged about his business acumen and fast dealing: "New York is a fast city. Its men are fast, its women are fast, and so are its horses."[5] Lyon complained that the proud carters lost their autonomy when blacks, Irishmen, and corporations (an unlikely combination) took over the trade in the 1850s, although the growth of express companies and hotel stages as rivals also cut into his business.

Teamsters

With the arrival of two-horse wagons in the 1850s and a great growth in business, urban horse controllers became redefined as teamsters (i.e., those who drove a two-horse team and the attached four-wheel wagon rather than a one-horse, two-wheel cart). Between 1850 and 1860, the number of teamsters in the United States more than doubled, while the number of carters declined. Clearly, teaming, a largely urban occupation, was growing rapidly and carting was declining. The increase became obvious between 1870 and 1900. In New York City the number of teamsters grew 311 percent, in Chicago 675.5 percent, in Philadelphia 350.7 percent, in St. Louis 243.8 percent, in Boston 412.6 percent, and in Baltimore 157.6 percent. In each case, the number of teamsters was growing at triple the rate of the human population.[6]

Teamsters were overwhelmingly male. Of the 120,000 teamsters in the United States in 1870, only 196 were women. Of the 368,000 in 1900, only 264 were women. In eighty years of census data for Boston, only one woman was counted. Women were stereotyped as too weak or too nervous to handle teams. One driving manual, while acknowledging that driving required "a gentle touch" and even recommending piano playing as good practice, complained about the woman driver that "she is fitted neither by garb, nature, nor habit to act." Large numbers of African Americans and Irish Americans worked as drivers. (By 1880 more than half of Boston's teamsters were Irish.) Blacks drove mules in many southern cities, and the *Encyclopedia of Southern Culture* notes that southern whites believed that blacks could read mules' thoughts. Even in a northern city like Pittsburgh, in 1911 more than half the teamsters were black. The Providence street railway assigned black horses to African Americans. The Irish, too, were believed to have "mysterious and telepathic" communication with horses. The real world, of course, was different. Both groups (blacks and Irish) came from rural areas where horse raising was common, so they had the appropriate skills. Blacks probably went into trades using horses because they had the relevant skills, they relished the opportunity to be more independent than in most jobs, and entry costs were low. The Irish background was probably similar.[7]

Often teamsters had started as child workers, handling light jobs as stable boys. One teamster complained about adolescent farm boys who would leave the farm after planting, use their horse skills to enter the teaming trade in cities—where they would compete with established drivers—and then return home for the harvest. Thus, teaming often served as an entry-level job for new migrants from the country to the city. Masculinity was an important element of self-definition for teamsters. One teamster leader wrote in 1912: "Because an individual is a teamster is nor [*sic*] reason why he can't be a man." But defining masculinity could be quite complex. Teamsters sought autonomy to be their own men, often leading to boisterous antiauthority behavior, cursing, drinking, and confrontations with bosses and the public. On the other hand, masculinity involved a patriarchal search for dignity and control. Dan Tobin, one of the first presidents of the International Brotherhood of Teamsters, saw no contradiction between his own heavy drinking and his writing of moralistic editorials in the union paper urging members to forego bad habits. Teamsters were often more interested in restraining arbitrary bosses through control over work rules and an arbitration process than in wages and hours, which was clearly a quest for dignity.[8]

The teaming trades required more skill than did carting. Carters usually controlled only one horse that they led on foot. Controlling two horses was much more complex. If one slacked, the other was overworked. Also, the outside horse worked harder in turns and was especially prone to shying or running away. In harnessing, it made sense to keep the "more nervous" horse to the right, away from traffic. The reins had to be held firmly enough to give the horses a sense of control but not so tightly as to make the horses insensitive to changes in pressure. Drivers could select from dozens of different types of bits, depending on the physiognomy and temperament of each horse. One manual described the choices as depending on the "lightness of drivers' hands" and "lightness of horses' mouth." Another gave seventeen different methods for holding the reins. Bit control was fundamental, but often all four reins were in one hand, leaving the other free for the whip. The prescriptive literature was inconsistent. Warnings were constant: "Always feel the horse's mouth," and "always keep a tight rein" (Wells Fargo). Yet a Chicago street railway manager warned drivers against excessive pulling of bit and reins, which could ruin a horse. He believed that horses should be controlled more by voice and whip. Another Chicago street railway operator wanted one hand on the reins and the other on the brake.[9]

Smooth handling was vital, since abrupt starts and stops could damage freight and passengers. The horses' legs are the weakest part of the equine machine. Poor driving could weaken and even permanently injure horses' legs. Surviving films of horse-dominated streets show drivers staring straight ahead (the strong point of human vision), leaving happenings to the side to horses, who have better peripheral vision.[10]

Learning to control larger teams was even more complex—four horses usually required a whip to ensure an equal workload, but that meant four sets or eight reins had to be held in the left hand, a skill that could take a year to acquire.[11] On rare occasions teamsters directed even longer teams, up to twenty-two animals, which required great skill, even with the aid of riders on lead horses. The City of Boston deemed four-horse teams so dangerous that using them required the permission of the Board of Aldermen.[12] The largest team we have seen mentioned involved forty horses moving nine thousand feet of reeled wire cable (weighing 12.5 tons) for a Chicago cable car system. The *Street Railway Journal* colorfully reported its start: "Forty whips were raised in mid-air; forty drivers uttered oaths of colossal proportions; forty horses felt the lash, and the North Side cable began to be threaded."[13] Large teams were also used to haul heavy loads in the iron and steel industry and to pull heavy stone monuments.

Controlling fire horses required special skill. They had to be trained to run to a spot under their harness when the bell rang, wait patiently for the harness to be dropped from the ceiling and hitched, speed to fires at high speeds, and then stand quietly and unattended by a curb while a fire raged nearby. Most fire engines required more than two horses because of their weight (often more than eight thousand pounds). Because they were top heavy, fire engines tended to turn over when corners were taken at high speed, and firemen would hitch three horses side by side, like a Russian troika, an arrangement that was very hard to control. Their drivers trained fire horses so well that, when they aged and were sold for new tasks, like pulling cabs, their new owners sometimes could not stop them from running back to their old fire house at the sound of a fire bell.[14]

Policemen had somewhat different problems. They were among the few individuals who rode horses in cities. They trained horses to be "well-bitted, bridle-wise and leg-wise" (i.e., steerable by leg pressure). Mounted police trained their horses to use six gaits, switch leads, angle off runaways, and slow down and push crowds sideways—a training process that could take six months. One famous Pittsburgh horse was even trained to hold arrestees' wrists in his mouth firmly enough to keep them from running but without breaking their skin. Both police and fire horses had to ignore loud noises and crowds.[15]

Braking, especially with heavy loads, also presented problems. Lank O'Dell, formerly a construction dray driver, drove the first street railway car in New York. He halted at its first downhill stop without a problem. From experience, he knew to manipulate both brake and reins. The driver behind him, formerly a cabbie, was not so fortunate. He forgot the brake and plowed into O'Dell's vehicle, an inauspicious start for the new mode. The Providence street railway company prohibited the application of brakes on hills, ordering drivers to rein in horses, despite the obvious stresses on horses' legs. The same road required that horses be taken out of harness during downhill runs in icy weather, expecting the driver to control the car by brakes alone. The first treatise on street railway management in the United States warned that drivers should dismount and walk their horses around curves and in front of schools, which suggests braking problems. Some street railways banned all stops on hills, fearing that downhill cars would overrun their horses and uphill cars would be unable to restart. Horses braked too rapidly often slipped and fell on slippery granite pavements. As late as 1931 many milk delivery companies (that trade was one of the last holdouts of the horse) had no brakes on their vehicles. Teamsters had to plan delivery routes to minimize wear and tear on their horses' legs—for example, delivering on the flat part of a route first to lessen the load. Training horses to back up to grates in sidewalks to deliver commodities like coal or beer kegs was even more complicated.[16]

Teamsters' responsibilities did not end with driving. Wells Fargo and other owners gave elaborate instructions about watering and feeding horses (most wagons carried some oats and a feedbag). The company asked drivers to avoid public water troughs (potential sources of disease) but did not suggest alternatives. Drivers had to deal with injured horses and get fallen ones to get up (put a blanket under its feet and "induce it to arise," said the Wells Fargo manual helpfully). Drivers also put blankets on horses in cold or rainy weather.[17]

Tracking the daily schedule of an ice driver shows some of the complexity of the job, beyond dealing with horses and traffic. Early in the morning, an ice man had to load four thousand to six thousand pounds of ice after estimating how much was needed (demand varied with the season and temperature) for delivery to 180 to 220 customers. He packed ice in blocks of one hundred pounds (smaller ones melted too fast) and then carved it into smaller blocks on the road (usually twenty-five to fifty pounds to fit in each customer's ice box—the exact size varied by the make of the ice box). Then the teamster drove the route, looking for cards that customers were supposed to leave in their windows when they needed more ice. A good deliveryman could often anticipate when customers needed ice, even if they forgot the card. The closer to each house the large block was carved the less the meltdown. The hotter the weather, the heavier the loads would be for both teamster and horse. Drivers also had all the problems associated with bill collection, since they were paid on commission. Companies started drivers with forty to fifty customers and took away routes if the iceman did not build it up to two hundred or so, a level that would generate a healthy income for the driver. Milkmen had to recruit customers, lift three thousand pounds a day, and be responsible for breakage and loss. Delivery horses suffered from heavy loads, frequent starts, and especially longer days in the summer.[18]

Teamsters took great pride in their skill. One proletarian novelist compared a teamster in action to "Apollo driving the sun across the sky."[19] As Theodore Dreiser described one of his teamster characters, "He drove as only a horse-lover can, his body bolt upright, his own energy and temperament animating his animals."[20] Control of self and animal were intertwined: "Horses show much cunning in alarming a timid rider," and some horses were "inherently vicious or wicked."[21] Horse authority James Garland warned about the need to concentrate, especially when there were objects or events likely to alarm a horse.[22] These traits were more related to the animal nature of the horse than to its machinelike features.

Some elements of the old carting system remained in place. New York had licenses as late as the 1850s, but nobody seems to have paid much attention to the requirement. In Providence the police commissioner estimated that the number of unlicensed drivers in the late 1860s was as great as the number with licenses. By the 1870s, licensed drivers, aside from hackmen, had disappeared, perhaps reflecting the reluctance of manufacturers and railroads to be dependent on independent drivers. Railroads wanted the lucrative local delivery business themselves. In 1858, for instance, Detroit's railroads took over local deliveries from draymen, who promptly organized, but the strike failed. By 1884 fewer than 10 percent of that city's teamsters owned their own rigs.[23]

The change in ownership patterns for horses and vehicles occurred elsewhere, too. In 1883, New York City teamster Thomas McGuire testified to a U.S. Senate Committee about conditions in the teaming trade. He had entered the trade as an independent, with capital of three hundred dollars, enough for one horse and wagon, but could not compete with the large firms, who used their purchasing power to reduce feeding and horseshoeing costs to a level one-third below his. The same complaint ran through the testimony of other drivers—competing with the big express companies like Wells Fargo was exceedingly difficult, and independents were only a dead horse away from bankruptcy. Even in the hack business, where it was harder to achieve economies of scale, corporate takeovers became increasingly commonplace as large firms began to monopolize licenses.[24]

The old independent cartmen and cabdrivers spent much of their time idling, waiting at stands or nearby saloons for business to come their way. Idling cartmen could be a nuisance, and cities tried to prevent them from swearing and smoking and ultimately banned the public stands. One Hartford resident complained: "I do protest against the stands on grounds of good citizenship, against the blasphemy and obscenity on the square. I have been insulted. Ladies have been insulted." In 1857, Milwaukee arrested thirty-five carters after the passage of a new ordinance that banned indecent or boisterous language or annoying travelers. The decline of independent operators ended the stands where carters had hung out.[25] Teamsters saw loss of status in these changes. There is little of Isaac Lyons's pride in his skills in a poem that a teamster named C. C. Hassler wrote for his union paper in 1902, in which he compared the toils and fatigue of horses and their teamsters, describing them as "fit co-partners in Life's race."[26] For Hassler there is none of the autonomy, craft, or skill that Lyons had. He views himself purely in working-class terms, sentimentally one with the "dumb brutes."

One of the first statistical surveys of occupational hazards in the United States pointed out that drivers and teamsters in New York City were exceptionally susceptible to pneumonia, rheumatism, and varicose veins because they worked such long hours, often in bad weather conditions.[27] Streetcar employees had an especially difficult work environment, with drivers putting in fourteen- and sometimes sixteen-hour shifts, even in the harsh winter months, working under onerous regulations, and having to play second fiddle to conductors. At least their work was regular, while teaming for other employers could be highly seasonal.

Teamster Organizations

Teamsters held an ambiguous place within the labor movement, as remains the case. Craft unions tended to reject teamsters, arguing that they were only semiskilled. Try as they might, they could never distinguish themselves by claiming that their abilities, including care with loads, promptness, interpersonal skills, and knowledge of geography, separated them from "farmers, who just knew how to drive horses." Their earliest unions were really guilds, which confined themselves to owners. The many unions that formed and then faltered in the last third of the nineteenth century spent much of their time arguing about whether to admit drivers who owned their own animals and rigs. There were other thorny questions. Should drivers be organized by stable? Should ice deliverers and laundry deliverers belong to the same union? How about cabbies? Street railway drivers? The American Federation of Labor argued that each was a separate craft. Indeed, the American Street Railway Union, confined to one trade, was probably the most successful early organization.[28]

Many drivers, such as Chicago's laundry drivers, drove their own horses and operated on commission, hustling for new customers, for example, at newsstands and hotels. Brewery drivers also could earn a high level of commission by bringing in new business. Even express truck drivers could earn money by finding new customers. Business owners with very few teams often welcomed citywide unions, which fixed prices as well as wages, thus regularizing competition. This often led to sweetheart deals, with attendant corruption.

The teamsters viewed their history as proletarianization. This was true in the sense that, in 1900, almost all were employees who did not own the tools of their trade—horses and wagons—unlike the carters who had dominated the carrying in 1850. There were, however, opportunities in the new order. In Chicago there were more laundry drivers with relatively high incomes than there had been teamsters and carters in 1850. The teaming industry seems to have been dominated by small firms that held contracts with manufacturing and distribution firms, rather than the large gangs of employees working for giant express companies and breweries. In this structure there was room for upward mobility, attended, to be sure, by the strong risk of failure. R. G. Dun and Company gave credit ratings for twelve hack or small teaming firms in Boston between 1858 and 1890. Four of them seem to have prospered and grown, while three clearly failed, two in the panic year of 1873. All were quite flexible, renting out their horses when idle and also renting stable space, a kind of livery/teaming combination. Of course, the vast majority of the city's teamsters and teaming firms had no credit ratings at all.[29]

Among larger firms teamster unions had their most success with breweries and department stores, which feared losing consumers if they were perceived as anti-union. Indeed, unions relied on public support in the form of boycotts of businesses employing scabs or even violence directed at scabs. Both employers and teamsters looked upon actual or expected violence as a matter of course. They had great success organizing firms that provided building supplies because people in the construction trades often would not cross picket lines of teamsters.[30]

The *International Teamster*, the official magazine of the teamsters union in New York City, noted in 1905 that threatening a strike in the food delivery business, after the trucks were loaded, was invariably successful, especially against ice cream companies.[31] A teaming firm could not lock its property up behind closed factory doors but had to operate it on the streets, vulnerable to the syndicalist instincts of many urbanites. Support by those in the street could be quite strong because of identification between customers and striking local drivers.[32]

Strikebreakers, however, seem to have been easy to find, suggesting that the skills of teamsters were not as unusual as they believed. New immigrants might not understand the cause of a strike, and blacks did not hesitate to break strikes of locals that would not allow them to join. In Chicago, black strikebreakers ended a strike of forty-six hundred teamsters that had lasted 108 days with several fatalities. Many of the early, failed unions seem to have a syndicalist twist, like the Liberty Dawn cabdrivers in New York, who affiliated with the Knights of Labor.[33]

The modern teamsters union, organized on industrial principles, dates back to Chicago in 1898. Its primary benefit may have been to reduce the hours of teamsters. Even though the horses rested on Sunday, teamsters had been required to come in and feed their own animals, a job the new contracts gave to stablehands, since teamsters were guaranteed double time for working on the Sabbath. Under the new contract milkmen made only one delivery run per day, reducing their workweek from eighty to fifty hours. Haulers had the workday cut from fourteen to eleven hours. There were some financial benefits. Railway express drivers, who acted as solicitors and were bonded as responsible, increased their income to seventy dollars a month.[34] For those hauling building materials, there was sometimes a loading/unloading piece rate in addition to the hourly wage (e.g., forty cents per one thousand bricks). The union freed livery and department store drivers from buying their own uniforms. Companies assumed the cost of theft, spoilage, and goods returned, if, for example, kerosene shipments spoiled groceries.[35]

By the end of the nineteenth century, teamsters were not only experiencing perceived proletarianization but also becoming stigmatized in new ways. The image of the rowdy, cursing, dishonest teamster, which went back to the times of the carters, remained. In 1895 the *Nation* described teamsters as "one of the lowest classes of the community"; they were also defined as foul-smelling, a kind of social pariah, just as hostlers were.[36] The sensationalist *New York by Sunlight and Gaslight* noted that "the odor of the horse blanket clings to them always."[37] The teamsters apparently internalized these fears—one of the characters of the proletarian novelist James T. Farrell bathed constantly and complained that he could never get the smell off his clothes, while another faced ridicule from neighbors because he smelled like "horse apples" (slang for manure).[38] Another proletarian novelist describes a teamster whose sister-in-law ridiculed him for spending his day "at a horse's tail." She complains that he spends all his time with other teamsters and lives on the top floor of a tenement in a space where only "goyim and paupers rented."[39]

Individuals stigmatized in this way create their own counterculture. The rowdy behaviors of drinking, contempt of authority, and camaraderie with fellows are classic responses. More formally, guilds, unions, and mutual benefit associations are common.[40] A description of Baltimore's "arabbers" (modern black street peddlers) catches this ambiance nicely: "Arabbers worked hard and partied hearty." The stable was a center of sociability, and one arabber says that he spent the happiest years of his childhood in a stable. The trade carried a stigma of hucksterism and "borderline vagrancy," which its practitioners valued, seeing one as a sign of business acumen and the other as a kind of republican autonomy. An arabber with a college degree was asked why he worked as a peddler. He responded: "I'll tell you the plain truth. In a word, freedom. I don't like kissing white people's ass."[41] It is not hard to imagine a white teamster a century earlier substituting class or ethnicity for these racial values.

Cruelty and Its Regulators

One of the scariest moments in American literature appears in Henry Roth's *Call It Sleep*. David, the novel's eleven-year-old protagonist, has been unable to prevent some drunks from stealing milk from his father's delivery wagon. When his always choleric father returns, he reaches for the whip. For one terrifying moment, David is sure that he will be the target of his father's rage. Instead, the old man takes his frustration out on the family horse. Nineteenth-century streets were filled with scenes like this—rage seems to have been a commonplace emotion, although one in decline, according to one history of anger. Horses, just like physically weaker family members, were perhaps victimized by teamsters who perceived status decline, by some excessively greedy owners, and even by wealthy, leisurely horse "lovers" seeking the perfect gait, the perfect posture, or the perfectly shaped tail. Even bystanders participated in the kind of sadism that the broader society sought to stop, such as children throwing firecrackers under horses or the young boy who put honey on a lamppost in subzero weather, knowing that a horse would lick it, freezing its tongue to the iron. As psychologists have demonstrated today, individual histories of violence against animals, especially by children, are often closely linked to later violence against people.[42]

One sign of decline in the acceptability of rage was the 1866 creation of the American Society for the Prevention of Cruelty to Animals (ASPCA) in New York City, an organization that spawned imitators in other cities. Its founder was the wealthy Manhattan socialite Henry Bergh; his father had made money in brothels and real estate. Bergh modeled the organization after anti-cruelty groups in London and Paris. These organizations clearly focused on horses—the ASPCA's *Annual Report* for 1884 showed that 90 percent of its prosecutions dealt with horses and especially emphasized controlling cruelty on public streets. Bergh and several of the other directors had been abolitionists before the Civil War and would later be involved in groups to prevent cruelty to children. The early directors were prominent members of New York City's elite society, many of them with investments in transportation firms with large herds.[43] As we shall see, their motives were more than just altruistic.

These individuals lobbied to secure a unique charter for the ASPCA, in which the state delegated police authority to make ASPCA agents deputy sheriffs with the power to arrest practitioners of cruelty. More importantly, and unlike the societies that followed in other cities, the ASPCA could keep any fines imposed on teamsters by the courts, a real incentive for active patrolling. Bergh even became an honorary prosecutor. Although he was not a lawyer, he sometimes did appear to argue prosecutions. This seems to have been a way around the refusal of officials sympathetic to teamsters to prosecute the anti-cruelty laws.[44] Bergh proved to be especially adept at generating publicity for his cause. He often arrested drivers himself and ordered overworked street railway horses returned to their barns, even though their passengers might be stranded in foul weather. He engaged in public controversies with important stable owners like street railway and cab company owner William Vanderbilt, even though he was a member of

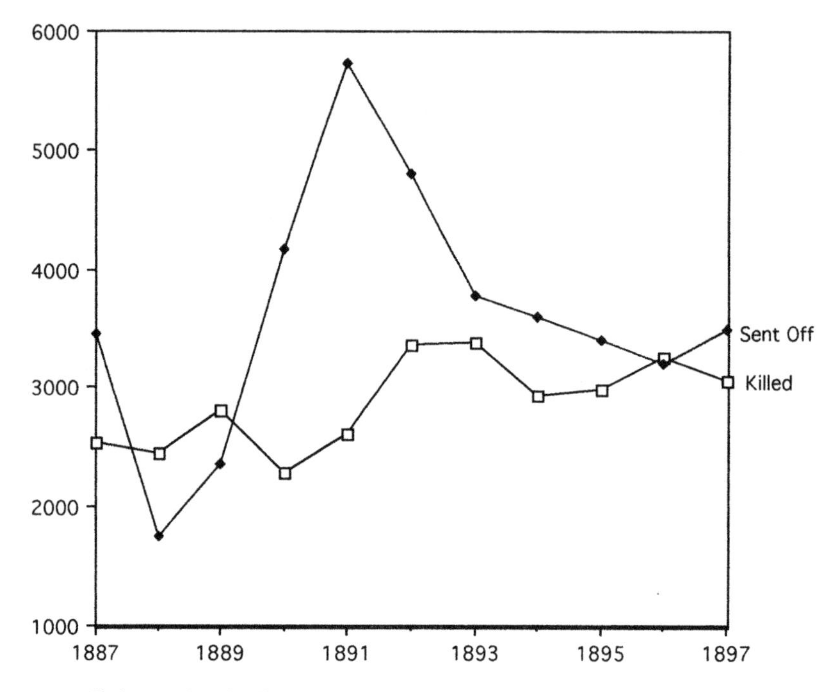

Horses Killed or Ordered Off the Streets by the ASPCA, New York City, 1887–1897. *Open squares,* killed; *solid diamonds,* sent off the streets.
Source: ASPCA Annual Report, 1897.

the ASPCA board.[45] As the graphs show, the organization became a major force on the streets, although it is hard to interpret the year-to-year variations, since they could well reflect enforcement zeal more than any objective change.

The 1884 *Annual Report* of the ASPCA summarized the reasons for prosecutions since its founding: harness sores, 3,538 (31.2%); beating, 1,513 (13.3%); lameness, 1,381 (12.1%); abandonment or starvation, 488 (4.3%); overloading, 611 (5.3%); working sick (mostly glandered) horses, 533 (4.7%); reckless driving, 257 (2.2%); maliciously killing or wounding, 281 (2.5%); selling diseased animals, 257 (2.2%); driving horses until dead, 12; exposing animals to storms, 138; and salting streets, 86. The report noted that many offenses occurred in bad weather, since rain or snow might exacerbate sores and increase falls. Bad weather caused many problems, such as overloaded streetcars and a higher than normal demand for cabs and for freight delivery even though footing was at its worst. One of the more aggravated cases in the ASPCA files involves a horse that fell in 1870 and was left lying on the ice, where its body warmth sank it five or six inches into the

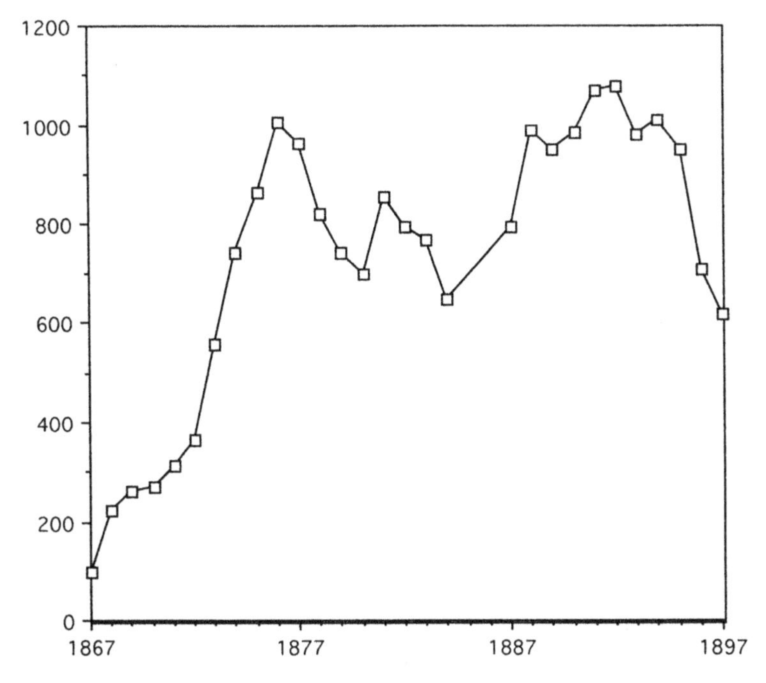

Cases Prosecuted by the ASPCA, New York City, 1867–1897.
Source: ASPCA Annual Reports.

ice overnight, requiring its destruction.[46] Summer had its own problems for horses, notably heat stroke.

Cruelty by teamsters toward their horses was undoubtedly often compelled by owners who insisted that teamsters hold to a tight schedule even though they were pulling excessive loads. Bergh acknowledged the unfairness of going after teamsters for $300 fines when their employers had insisted on overloads, but he found that he had little success prosecuting the latter. Street railway owners tried to remain on cordial terms with Bergh, although they (at least by Bergh's account) blocked legislation that would have limited the number of passengers that their horses could pull and mandated extra horses on steep hills. When a January 5, 1887, letter to the *New York World* complained that horses had been overworked during a recent snowstorm, the ASPCA responded that the president, directors, and officers of the offending street railway were all ASPCA members: "They certainly take pride in being members for it gives them the power and authority to stop all new cruelty." One newspaper reported an extremely gross beating but noted that Bergh remonstrated with but did not arrest the owner of a carriage and "fine span of horses" near Central Park. Bergh probably did not like to prosecute members of his own class.[47]

The pattern in Pittsburgh was slightly different. The Allegheny County Humane Society (ACHS) reported a few prosecutions for using horses without shoes (their owners must have been close to bankruptcy, since this practice would clearly ruin horses). In 1876 that agency reported sixteen hundred cases of cruelty. It never had the power to prosecute of the New York ASPCA, and almost all stops led to some form of counseling, with few fines and fewer arrests. After being reconstituted as part of the Pennsylvania SPCA in 1880, the organization announced that its goal was "to prevent, not punish." Still, the Pittsburgh society was willing to take on wealthy horse owners who used cruel techniques and won a major court case in 1889, with the ruling that docking and check reins (reins on trotting horses that held their heads in an artificially high position, a fashionable look) were inherently cruel.[48] Most local humane societies, however, did not seem to go after inherently cruel forms of horse fashion in the upper-class world, such as docking (cutting tails short for cosmetic reasons and thereby limiting the ability of horses to swat flies), playing polo, and show jumping. (Both polo and jumping stress the legs.)

Arrested teamsters argued that owners should be prosecuted, not them, usually claiming that the owners had given them ill-fed or lame horses and insisted that they extract work from them. One teamster told Bergh, "I licked him [the horse] because he was unable to stand up and the company has no mercy on either the driver or the horses—and I must make time."[49] Drivers for firms demanding exact on-time delivery often had no recourse but heavy whipping or other cruel practices, such as shoving a sharp stick up the animal's anus, twisting his tail, or starting a fire underneath him.[50] One teamster quoted his boss: "God damn you, buy whips and I will buy horses. When he is dead, I will give you another one."[51] This statement clearly reflected a view of the horse as a machine rather than as a sentient being.

Perhaps the best measure of how complicated these issues could be revolved around street railways in bad weather. Bergh would often stop cars, especially on hills, until extra horses were added. But this could cause monstrous jams, with up to fifty cars backed up behind the one without horses. At first he sought prosecutions directed at drivers but, realizing that they often were not responsible, he then went after the owners, but often with a lack of vigor. The owners protested and often had the clout to avoid prosecution. Finally, Bergh decided to blame passengers for overcrowding (sometimes with good reason—his agent found 137 riders on one 40-passenger car during a blizzard in 1873). He forced the men and boys (but not the women) on the car to debark and walk up a hill. In 1887 an ASPCA spokesperson actually told the *New York World*, "The people who crowded the cars were the real brutes."[52]

Here and in other policies the ASPCA often served the interests of large herd owners. For example, Bergh's agents would shoot lame horses, making it easier for owners to file claims with insurance companies, since the company would refuse to pay if the beneficiary had shot the animal himself. (Ironically, the society soon became the leading killer of horses in New York City.) Agents also shot infected animals before they could spread disease, notably those with the highly infectious glanders. Individual stable keepers, especially those who owned only one or two horses and could not afford a replacement, might acknowledge the need for such public health measures, but few would actually allow their own animals to be shot.

Nor did most owners want their stock abused by their human employees. Wells Fargo, for example, prohibited whips. Other firms clearly limited whipping in their regulations.[53] In 1919, an official of the National Biscuit Company told the Allegheny County Humane Society: "We have a good deal of money invested in horse flesh and we surely appreciate the cooperation of the Humane Society in helping us see that our horses are handled in a humane manner in the street."[54] Companies like those could hardly have had inspectors everywhere on the streets, so organizations such as the ASPCA and the ACHS helped protect their valuable animals by enforcing anti-cruelty laws. In a sense, the owners were capitalizing on Victorian sentimentality—drivers could not beat animals out of sight of the owners. Absentee stable owners probably welcomed ASPCA inspections. The Massachusetts Society for the Prevention of Cruelty to Animals may have best summarized the relations between employers and drivers: "If the men are on good terms with the employer, the horses are usually well treated, whereas if the men are dissatisfied, the horses are always badly treated."[55]

New York's street railways initially stopped whenever a prospective passenger flagged them down, a practice borrowed from their predecessor, the omnibus. In 1874 the companies switched to stopping only at street corners, still the dominant pattern in American transit, because it saved wear and tear on equine legs. The street railways forestalled consumer protest by claiming endorsement by the ASPCA—frequent stops represented cruelty to horses. The ASPCA ran an educational campaign encouraging New Yorkers to board cars at the corners only.[56] Reducing stops also improved travel times, a goal of the street railways.

Note the ambiguity here. While anti-cruelty groups were ostensibly focused on the living side of living machines, many of their policies also served to facilitate the use of horses as machines. The reality never quite matched the rhetoric and, of course, it introduced the familiar paradox of humans probably being the leading killers of animals in cities.

Early in the twentieth century, anti-cruelty crusaders turned to more positive reinforcement, holding annual workhorse parades in cities across the United States. These parades became major civic events, with thousands of spectators. Individual teamsters displayed their animals, often adorned with straw hats to protect them from the heat, and were rewarded for the healthy appearance of their animals. After Boston's 1919 parade, the president of the dairy firm H. P. Hood estimated that the parade saved him one thousand dollars a year because of the extra care stable hands and drivers took of horses, hoping for a ribbon. *Outing* magazine reported, "Another employer, a shrewd, successful Jew owning 14 horses . . . reports that his horses have increased in value $25 or $30 apiece as a direct result of the stimulus afforded by the parades."[57]

Teamster-based sources show a different attitude toward cruelty than that attributed to them by the middle class. One Providence street railway driver refused to whip a stopped horse, even though eleven cars were backed up behind him. Eventually his passengers pushed the vehicle back to the barn. Seattle teamsters included as a strike demand a request that horse feedings be increased from two to three a day. Teamster newspapers praised anti-cruelty crusaders despite the obvious tensions between the groups and also warned against excessive whipping. In 1906 the *International Teamster* suggested to members that they should gently break horses of the habit of shying and also not trot them downhill. The same magazine carried photos of especially handsome teams in almost every issue. In the middle of a bitter and violent strike in New York City in 1887, members of the Liberty Dawn union chased a scab from his vehicle and overturned it but walked the horse back to the stable.[58]

Objectively, how much cruelty was there? The constant recurrence of complaints about it suggests that it was commonplace. After all, there was formal legislation, pressure from quasi-private anti-cruelty groups, rules by employers, and even admonitions from teamster groups. Those implicated (except socialites who used check reins, docking, and the like) never defended abuse per se, other than citing the exigency of the moment, such as the need to deliver goods on schedule or to train a horse properly. Even accepted practices were inherently cruel—yanking on bits hurt, braking did pull horses to excessively sharp stops, and towing weights of any kind could be hard. Very few drivers, if any, abandoned the whip. So some cruelty was needed to make cities work. And humane societies needed cruelty to justify their existence—indeed, there is the great irony of their being the leading killers of horses.

Corporate and Municipal Control

Urban work requirements created other complex issues of human-horse interaction, especially in large corporate herds. Since drivers controlled their horses outside the owner's direct view, owners worried about cruelty or neglect that would damage their living machines. Thus, the management of teamsters, and not just by the threat of legal prosecution, was also important. One company charged employees "to know every horse as a mother knows her child."[59] Once again, the direct simile is a reference to humans. Another firm reminded drivers, "Don't forget he is a sentient being and can feel the lash of either your whip or your tongue."[60] At some level managers understood that they were dealing with two species of sentient beings. One teamster recounts being told, on his first day on the job, "Animals are a lot like people. Each of these horses has a different personality, but all of them need tender, loving care. . . . Animals always remember the people who treat them right."[61] At its annual banquet the American Street Railway Association toasted, "Mankind's noblest servant, co-worker, and friend: the horse."[62] Drivers could be equally sentimental, as in the poem quoted previously.

Owners tried to match horses and drivers by personality or other traits to create bonds of affection between both species of workers. They encouraged drivers to name their horses, humanizing them. Boston's workhorses included Houyhnhnm (an especially flattering name, derived from the horses that were smarter than humans in Swift's *Gulliver's Travels*), Grief, John Wesley (the famous Methodist preacher), Foreordination, Hayes (a president), Tilden (a failed presidential candidate), Richard III (after the English king whom Shakespeare portrayed as dying in battle for want of a horse), Major, Jumbo (after the famous circus elephant), and Russell Sage (a well-known millionaire). All this suggests that owners and teamsters sometimes perceived their horses as companions, as well as machines. This kind of bonding had practical benefits in traffic. If twenty drivers cried conflicting orders on a crowded street, it was important that each horse distinguish his driver from others. Commands to horses could vary with each teamster-horse pair, both for safety and, one suspects, bonding. One immigrant teamster told his grandchild that his horse understood Yiddish. Another had horses, Giovanni and Giuseppe, who obeyed only commands in Italian. The bonding probably worked both ways. We have not encountered an account of a teamster strike where strikers mistreated horses, although they rarely hesitated to sabotage other property. In one case, an especially bitter street railway strike, teamsters even allowed feed wagons to cross their picket lines into the stable. Photos of early-twentieth-century workhorse parades show horses whose teamsters had decorated them with straw hats and ribbons in their manes. There were limits—while most firms asked their drivers to avoid the whip, it was still routine equipment. As late as 1912 one stable management manual recommended lighting a fire under horses to get them started with heavy loads.[63]

Training horses for city life was a difficult and important part of controlling them. The evolutionary track taken by horses provided shyness and speed as defense mechanisms. Horses scare easily, and their reflex is to run away. An overly excitable horse might cause an accident or damage itself if it reverted to this behavior. The first instruction on driving in one manual was: "Watch for objects likely to alarm your horse." Even a flying piece of paper can scare a horse, especially since their instinct is to watch the scenery to the side (to help navigate a return), rather than the road. Other handbooks recommended singing or whistling to horses to keep them calm. Accidents were common, especially in traffic. Possibly New York's earliest traffic jam involved the numerous carriages traveling to the famous match race between Eclipse and Sir Henry in 1823, the greatest equestrian event in the city's history. The heavy traffic led to a chain reaction accident, during which many panicky horses drove their traces through the rear panel of the preceding vehicle. Per vehicle, nineteenth-century horse-drawn vehicles caused more accidents than motor vehicles would later, an appalling accident toll, at least in New York City.[64]

Horses on the street were subject to formal municipal control, but regulation seems to have played little effective role. There were speed limits, but those were nearly impossible to enforce, since there was no effective way to measure speed. Boston, to cite one example, passed speed regulations at least four times in the nineteenth century, a sign that the old ones had been forgotten, presumably because of nonenforcement. In any case, most street traffic proceeded at a walk, to judge from films taken around the turn of the century.

Common law prosecutions for "furious driving," while rare, were probably more successful than those for violating speed limits. Furious driving could be demonstrated in a way that a speed limit could not—its measure was that horses frothed at the mouth. There were other rules, but they seemed mostly to affirm custom or common sense—keeping to the right, for example, or not leaving horses unattended at the curb. Turn-of-the-century film clips show no example of either custom being violated. Indeed, those clips are nearly devoid of any regulatory presence—there are no signs or signals, and only one traffic police officer is shown. To be sure, New York City claimed by that date that it had a police officer stationed at every corner of Broadway, its busiest street and probably the busiest in the nation, but none is visible in the film clips of that street. The films show that, while Broadway was busy, there were plenty of gaps in traffic and no vehicle on a cross street had to hesitate before turning into the flow of traffic. A traffic count of ten of the busiest streets in the nation in 1886 showed on average 108 vehicles an hour on Broadway. Only one other major street in the ten cities surveyed carried more than 75 vehicles per hour. Evidently there was some rush hour crowding at bridges, ferries, and major downtown intersections. For control of traffic, custom, as noted above, was clearly more important than regulation.[65]

Hacks (four wheelers) or cabs (two wheelers), which functioned like today's taxicabs, were the great exceptions to the lack of regulation. John Clapp received New York's first hack license in 1692. The arrival of the horsecar did not drive hacks out of business as their drivers had feared, since they served a separate, luxury business. By 1862 the numbers of horsecar and hack drivers in New York City were roughly the same—just over five hundred of each. Of course, the horsecars carried far more passengers. With considerable resistance from hack owners, a few large East Coast cities also adopted lighter, less expensive two-wheeled cabs (two-wheeled vehicles were cheaper because they required only one horse) after New York City licensed the first one in 1886. New York's ordinance allowed hacks to charge $1.25 a ride and cabs 75 cents. There were 945 cabs in New York City by 1866 and perhaps double that at the end of the century. Boston allowed cabs to charge 25 cents to $2.00 on a zone fare system and hacks twice as much. Either form of service tended to use the cheapest, often the oldest and most run-down animals. Other cities imitated New York. Regulations did not vary much from city to city, concentrating on fixing rates and licensing both drivers and vehicles, in part for purposes of taxation. Intricate issues were involved, especially the setting of rates, conflicts between owners and drivers, and tax policy, and deserve further examination. The ten cities that reported on hack regulation in the 1890 census reported roughly one cab for every one thousand people, although there was great variation from city to city.[66] While cabs and hacks were certainly part of the horse-drawn urban traffic scene, as on-demand vehicles they had a very limited and almost impossible to measure impact on the structure of the city and on urban travel behavior, the subject of the next chapter.

Horses, whether viewed as living machines or as sentient beings, became objects of control for workers, owners, philanthropists, and municipalities in the increasingly horse-dependent nineteenth-century city. These different groups became stakeholders in the use and treatment of the urban horse but often from different perspectives and using different methods. The most important issues related to the use of the horse to haul freight and people and what constituted exploitation of both the living machine and the teamsters who drove them. While cities did have some ordinances regulating horse use, they largely privatized the function by the granting of law enforcement powers to private agencies in the form of humane societies. Various patterns of cooperation and conflict were later reflected in other issues such as stable construction and maintenance and the regulation and operation of street railways. These changes required additional control in both the private and public sectors.

Milk delivery by a horse-drawn sled in Rochester, 1860.
Source: Courtesy of the Rochester Public Library.

A horsecar laboring up Fifth Avenue in Pittsburgh, c. 1865.
Source: Used with permission, Carnegie Library of Pittsburgh.

No streetcar lines were permitted on Broadway, so only omnibuses and cabs are visible in this Broadway street scene, 1867.
Source: Author's collection.

The famous Bull's Head Horse Market, New York City, 1869.
Source: Harper's Weekly, Feb. 13, 1869.

A combination livery stable and undertaking establishment in the East Liberty neighborhood of Pittsburgh, c. 1870. Notice the hearses on each side of the stable. Several Pittsburgh livery stables doubled as undertaking establishments.
Source: Used with permission, Carnegie Library of Pittsburgh.

A horse-powered ferry in Montreal harbor, 1870. Note the treadmill on the boat's deck. Horse-powered boats were commonly used in American cities before steam power, and a few continued to be used into the twentieth century.
Source: MP-1984.47.70. Used with permission, Notman Photographic Archives, McCord Museum of Canadian History, Montreal.

"Throwing Dead Horses into the Harbor of New York at Night." Horse carcasses, stripped of their skins, were thrown into the Narrows around Coney Island rather than being dumped into the Atlantic Ocean. The magazine reported that there were fifty to one hundred carcasses of different animals floating about or "roasting upon the shore" from Fulton Ferry to Coney Island.
Source: Frank Leslie's Illustrated Newspaper, Aug. 20, 1870.

Children playing near a dead horse lying on the street, c. 1900. Because dead horses had value, the carcass probably did not remain on the street for very long.
Source: Author's collection.

A Pittsburgh omnibus, c. 1840s. The Citizens Passenger Railway operated omnibuses on Penn Avenue in Pittsburgh.
Source: Author's collection.

"How Horses Are Abused," from a sketch by Thomas Worth. In the *top left,* a driver is urged to use the whip; *top right,* a "balky horse" is beaten with a stick ("Paddy's method"); *center bottom,* three policemen show "the majesty of the law"; *bottom left,* a boy warns his dad, "Here comes Bergh," the founder of the ASPCA.
Source: Harper's Weekly, March 27, 1880.

Traffic jams often occurred in New York City. Note the runners on the sleigh *(center)*, the Fulton Ferry omnibus *(right)*, the angry drivers, the unruly horses, and the horse carcass in the street in this satirical cartoon drawn by Gray-Parker.
Source: Harper's Weekly, Dec. 29, 1883.

The elite class promenading in Central Park, 1883.
Source: Harper's Weekly, May 19, 1883.

Jacques Cartier Square market with freight sleds, Montreal, 1884. Note the hay for the horses, placed on the ground.
Source: VIEW-1487. Used with permission, Notman Photographic Archives, McCord Museum of Canadian History, Montreal.

Industry used horses for multiple functions. A two-wheeled cart is much easier to unload than one with four wheels. Pittsburgh, c. 1890.
Source: Used with permission, Carnegie Library of Pittsburgh.

Without consistent and thorough manure removal by scavengers or street cleaners, city streets could look like New York's Morton Street on March 17, 1893.
Source: George E. Waring Jr., *Street-Cleaning and the Disposal of a City's Wastes* (New York, 1898).

A New York City street sweeper in 1896. This sweeper was obviously resistant to reformer Col. George E. Waring's attempt to dress all members of the Street Cleaning Department in white duck uniforms!
Source: New York Public Library, ID 79772.

Laundry delivery had to continue even in the winter. Photograph c. 1900.
Source: Courtesy of The Albert R. Stone Negative Collection, "Schuyler Townson" Library, Rochester Museum & Science Center, Rochester, New York.

Workhorse parade, Rochester, c. 1900. Such parades were common in cities during the late nineteenth and early twentieth centuries.
Source: Courtesy of The Albert R. Stone Negative Collection, "Schuyler Townson" Library, Rochester Museum & Science Center, Rochester, New York.

Accidents were common, especially on slippery streets. This downed horse had slipped on a patch of ice, a major occupational hazard in northern cities. Photograph c. 1900. *Source:* Courtesy of The Albert R. Stone Negative Collection, Rochester Museum & Science Center, Rochester, New York.

Horses were critical in supplying city services. Here, a four-horse engine races to a fire, followed by a two-horse team with more firemen (Minneapolis, c. 1900). *Source:* Used with permission of the Minneapolis Public Library.

Rolls of paper for the *Chicago Daily News* will be unloaded from a horse-drawn truck in 1903.
Source: DN-0001448, Chicago Daily News negatives collection, Chicago Historical Society.

A prize three-horse team delivers Heinz food products in Pittsburgh in 1904.
Source: Used with permission, Library and Archives Division, Historical Society of Western Pennsylvania, Pittsburgh, Pennsylvania.

Traffic on Liberty Avenue, Pittsburgh, 1906. Note the mixture of vehicles on the street, including horse-drawn trucks and carriages, motor-driven trolleys, and automobiles. The number of automobiles was continuing to increase, creating major traffic problems in the city's downtown.
Source: Used with permission, Library and Archives Division, Historical Society of Western Pennsylvania, Pittsburgh, Pennsylvania.

A Molson's Brewery beer cart, Montreal, 1908. Horse-pulled beer wagons have persisted into our own time as a cultural icon.
Source: VIEW-8752. Used with permission, Notman Photographic Archives, McCord Museum of Canadian History, Montreal.

Mounted policemen help protect the public safety and also serve for crowd control. Police inspectors line up for the Pittsburgh Sesqui-Centennial Parade in 1908.
Source: W. H. Stevenson et al., eds., *The Story of the Sesqui-Centennial Celebration of Pittsburgh* (Pittsburgh, 1910).

In 1911, a horse drinks from a watering fountain in Chicago while a Fountain Girl watches and other horses wait their turn. Cups for humans hang from a chain attached to the statue's arm. Cities eventually eliminated fountains and troughs like this because they spread infection.
Source: DN-0056947, Chicago Daily News negatives collection, Chicago Historical Society.

A horse-drawn converted omnibus picks up students at the Fallow School in Chicago, 1911.
Source: DN-008953, Chicago Daily News negatives collection, Chicago Historical Society.

Winter horse racing on Tremont Street, Rochester, c. 1920. Even the urban horse could be an adventure machine for both participants and spectators. Note the Great Atlantic & Pacific Tea Company *(right)*, later the A&P.
Source: Courtesy of The Albert R. Stone Negative Collection, "Schuyler Townson" Library, Rochester Museum & Science Center, Rochester, New York.

Hay Market, Rochester, c. 1920. By this date, farmers usually baled their hay to reduce its bulk.
Source: Courtesy of The Albert R. Stone Negative Collection, "Schuyler Townson" Library, Rochester Museum & Science Center, Rochester, New York.

Children enjoying a shower from a horse-drawn sprinkler truck in Chicago, 1927.
Source: DN-0083661, Chicago Daily News negatives collection, Chicago Historical Society.

Snow removal, Rochester, c. 1935. As late as this date, many public works departments still relied on muscle, both equine and human.
Source: Rochester Municipal Archives.

A massive traffic jam in the Chicago Loop, c. 1905. Notice the mix of horse-drawn wagons, mounted policemen, and electric trolleys.
Source: Chicago Historical Society.

CHAPTER THREE

Powering Urban Transit

From 1830 to 1860 American urban areas experienced a "mobility revolution" driven by a shift to new transportation technologies. These technologies included horse-pulled public transportation, such as omnibuses and street railways, and the steam railroad. They were revolutionary changes because they combined "distance and regularity"—as urban historian Henry Binford notes, the "exceptional" trip became "ordinary," and the ordinary trip grew in distance.[1] This chapter focuses on the horse/vehicle driving machine and how it stimulated the evolution of the modern American city—a city characterized by a central business district, specialized residential neighborhoods, and peripheral suburbs. The development of horse-drawn public transit had more than geographic effects, however, and included social, cultural, gender, and political conflicts over the new technology. The evolving modern city was increasingly segregated by function and by social class. Travel itself in the horse-drawn vehicle came to reflect the tensions and prejudices of urban society.

Technologically, the key to taking advantage of equine mobility was to reduce fares by minimizing the number of prime movers. We have already explored one important element of this—the development of larger, stronger horses. In the mid-nineteenth century the two other important elements of mobility technology—the creation of smooth operating surfaces and reductions in vehicle weight—also improved. With smoother operating surfaces horses had less friction to overcome in pulling vehicles, while reduced weight meant that more of a horse's strength was used for the payload. Lower fares meant more mobility, whether suburban commuting by the middle class, resort travel by inner-city residents, or journeys to the central business district for work, shopping, or entertainment. The three criteria of reasonable price, fixed schedules, and predetermined routes were critical to the evolution of urban transit systems. Hacks and cabs, discussed in the previous chapter, were also important urban horse-drawn vehicles, but they did not have fixed schedules and responded, like the modern taxicab, on demand. Their constraints—the lack of a fixed schedule and route and the charge of relatively high fee—meant that they had a relatively smaller effect on the structure of the evolving city than did other horse-powered vehicles, such as the omnibus and especially the streetcar.

The Omnibus

The horse-drawn omnibus, which evolved from the traditional stagecoach, was the transit technology that satisfied the two criteria of fixed schedules and predetermined routes. Entrepreneurs first introduced the vehicle in France in the mid-1820s. Parisian police had initially resisted the innovation because they feared that omnibuses would create public safety issues by mixing various classes and would also obstruct narrow streets. Authorities in the French cities of Nantes, Bordeaux, and Lyon were less concerned about these issues, and the omnibus first appeared there.[2] In 1828, finally convinced of their safety, the Parisian police approved them. The prefect of police required the omnibuses to seat no fewer than twelve and no more than twenty riders along fixed routes. The first company to begin operations had vehicles, pulled by three horses, that seated a total of fourteen passengers in three differently priced compartments. The omnibus was an instant success with the Parisian bourgeoisie. By the end of 1829, ten firms operated a total of 264 coaches; ten years later, thirty-five lines operated 409 coaches. In 1854, just before consolidation of the omnibus lines into a government-sponsored monopoly known as the Compagnie Generale des Omnibus (CGO), 34 million passengers rode the omnibus lines.[3]

Crowded London also needed better public transport. Beginning in the first years of the nineteenth century, short-distance stagecoach routes bringing passengers from the suburbs to the city became increasingly popular. In the city itself, however, a hackney monopoly blocked Parliament from granting permission to run omnibuses regularly along city streets until 1832, when Parliament gave its approval, and franchised omnibuses, described as little more than boxes on wheels, finally began operating on London streets. London's omnibuses were initially bigger than the Paris omnibuses, seating twenty passengers behind three horses. Later London omnibuses were smaller to better navigate narrow city streets, with two horses pulling vehicles that carried fourteen passengers. By 1838, 620 omnibuses operated in the city.[4]

Although not yet as large as Paris or London, major cities in America, such as New York, Boston, and Philadelphia, also required improved public transport. As in London, regularly scheduled short-distance stages from suburban areas preceded the omnibus. Such stages began operating in the New York area as early as 1811, and by 1816 they were running every two hours to various suburban towns. By the mid-1820s stagecoaches were clogging Broadway and raising concerns in the Common Council about traffic regulations and license fees. These coaches, however, kept irregular schedules, charged very high fares, and were quite uncomfortable.[5]

In 1827 Abram Brower began operating an intracity stagecoach on Broadway in New York City. Two years later, he introduced a vehicle that had rear entry and lengthwise seats, setting the pattern that would be followed for all transit vehicles until the twentieth century. Unlike the Paris omnibus, however, it sat high off the ground. In 1830, supposedly inspired by a drawing of a low-slung Parisian city coach, Brower had his coach maker, John Stephenson, imitate the Parisian omnibus. Stephenson innovated by adding elliptical springs for a more comfortable ride and easier pull for the horses. Because New York's paving was so rough, the city's small omnibuses required three and four horses. The omnibuses charged a much lower fare than did stagecoaches, initially twelve cents, and followed fixed schedules on predetermined routes.[6]

By 1833 "the age of the omnibus" had arrived in New York, and the *Gazette and General Advertiser* entitled Gotham "The City of Omnibuses," since the municipality had issued licenses for eighty vehicles. In 1839 an observer counted sixty-seven omnibuses passing a point on Broadway during a half-hour afternoon period. An 1853 guidebook reported that omnibuses averaged 13,420 trips a day and collected 120,000 passenger fares. In 1848, 327 omnibuses were licensed, and in 1851, 568, with the highest number, 683, reached in 1853.[7]

The omnibus diffused rapidly from New York through the American urban network, with most large cities—those with populations over forty thousand—adopting them in the 1830s. In 1833 an omnibus line began running in Philadelphia on an hourly basis for 12½ cents, with later lines selling annual tickets for frequent riders. By 1848 the city had eighteen separate omnibus lines and about 130 omnibuses. Six years later 320 omnibuses ran on thirty different routes.[8] An omnibus first appeared in Boston in 1834, and by 1840 eighteen omnibus lines were operating; by 1847, 250 omnibuses and stages ran in the Boston area. About half the omnibus traffic went to Roxbury or other neighboring towns such as Charlestown, which had fifteen-minute service, and Cambridge, which had half-hour service.[9]

Transit entrepreneurs began operating omnibus lines during the 1840s in smaller cities such as Albany, Cincinnati, Lowell, and Pittsburgh, followed in the 1850s by Louisville, Newark, Providence, Milwaukee, and Rochester. The prerequisite for urban development of an omnibus line was probably the existence of a substantial middle- and upper-class population that sought to live in a city with segregated residential and business districts—a condition mainly made possible by internal urban public transport. One student of the omnibus notes that its adoption was "synonymous with a city's achievement of urbanity, and industrial urban structure."[10]

The typical route configuration grew out of those of earlier suburban stages and resembled spokes on a wheel. Omnibuses initially left hourly and took approximately one hour to reach a destination up to five miles from the center city. The second type of omnibus service evolved from hackneys and often began at railroad terminals on the periphery of the central city, from which passengers required transportation to hotels and commercial activities in the core. As railroad passenger volume grew, regularly scheduled omnibuses began operating from the stations. Over time, the two patterns tended to overlap, and lines developed along major thoroughfares. Only very large cities like New York had crosstown lines.[11]

Four horses usually pulled American omnibuses, although those seating more passengers occasionally required more. New York omnibuses weighed about twenty-five hundred pounds, and the horses weighed between one thousand and eleven hundred pounds.[12] In smaller or hillier cities, vehicles were lighter and might be pulled by only two horses. Omnibuses usually contained twelve to twenty-eight seats, with riders sometimes sitting on the roof. Crowding was common—one 1841 commentator observed that "compression is the principle."[13] The cars operated at two to five miles per hour, depending on road conditions and topography. In most cities speed limits existed on specific streets, but measurement was difficult. The Brooklyn Common Council, for instance, enacted a 4-mph speed limit on a section of busy Third Avenue in 1850, with a speeding penalty of ten dollars.[14]

Omnibuses were either individually owned or operated by larger companies, although over time they consolidated, especially in response to economic downturns, such as the crash of 1837. In 1852, for instance, when most New York lines were concentrated in lower Manhattan south of Forty-second Street, approximately thirty lines operated more than seven hundred omnibuses. Initially they stopped for passengers when flagged down, and competition between drivers resulted in races to pick up passengers and omnibus "wars." A new city charter approved in 1839 required omnibus and stage licenses, and a special act of the Common Council passed in 1854 regulated routes. Similar omnibus regulations were passed in Brooklyn in the 1840s after an omnibus war there between drivers competing for passengers.[15]

Omnibus passengers frequently complained about bumpy rides and vehicles that surged from "side to side." Streets paved with cobblestones provided more traction for horses but at the cost of passenger comfort. Macadam roads were preferable, but the ride could still be uncomfortable.[16] Rough streets necessitated frequent omnibus repairs or replacement.[17] Sidney George Fisher, a Philadelphia diarist, described the omnibus as "heavy, jolting, slow and uncomfortable." The *New York Tribune* commented, "The arrangements [of the omnibus] for shooting passengers out into the mud are unsurpassed." The *New York Herald* described the omnibus ride as "a perfect Bedlam on wheels." New York pickpockets were quick to take advantage of the opportunities presented by crowding, increasing the ride's hazards. Still, until confronted by streetcar competition, ridership continued to increase.[18]

Headways varied from five to fifteen minutes on the busiest streets. "You pick your teeth on the Astor House [New York] steps," wrote journalist George G. Foster, "and see, on an average, fifteen omnibuses pass each way, every minute, and for the great part of the day, all full."[19] In New York, which had the most extensive omnibus system in the nation, the cars moved from uptown avenues to a common destination on Broadway at City Hall Park, usually at a frequency of every five minutes. Massive congestion often developed at this point, as omnibuses merged with carts and hacks as well as with private vehicles, creating traffic "chaos."[20]

Omnibuses began regular service in Pittsburgh in the 1840s, and by the 1850s the city had four lines. These ran from the center city to contiguous towns such as Lawrenceville, Minersville, Oakland, and Allegheny City, following earlier coach routes. The largest line, the Excelsior, ran buses every ten minutes from its downtown depot to Lawrenceville, on the Allegheny River (a distance of about three miles) until ten in the evening for a fare of twelve cents. Competition between omnibus lines for space in the heart of the city was so strong that, in the early 1850s, the Pittsburgh City Councils passed an ordinance establishing an omnibus depot and requiring the vehicles to be "ranged in close order along the curbstone" while waiting for passengers.[21]

Because it was a public vehicle, the omnibus often evoked comments on its democratic character. In France several nineteenth-century commentators observed that the omnibus was "the sanctuary of equality" and that "everyone" passed through it, making the "history of the omnibus . . . a history of society."

In London one author touted the omnibus as existing "for the convenience of the many" and "shuddered at the thought of depriving a poor man of his omnibus."[22]

The view of American observers was the same. A New York commentator noted that different classes rode the omnibus at certain times during the day: "In the early morning . . . the omnibus is chiefly occupied by junior clerks with big iron keys in their hands, or laborers with tin kettles between their feet, on their way to their downtown avocations." Later in the morning came "sleek and rotund burghers," followed by "Gotham's fair wives and daughters" on their way to shop or to see their dressmakers. And the evening would find "young gentlemen" on the way to the theater.[23] Social mixing was distasteful to some, who felt that the omnibus lacked "distinction and taste," often carrying riders who were drunk or unclean. The omnibus, complained an 1841 writer, was a "parable for the evils of democratization" at a time when New York society was becoming more heterogeneous.[24]

Later commentators, however, noted the limited working-class ridership of the omnibus. Omnibus fares in New York, Brooklyn, and Boston were initially twelve cents, although they were cut in the face of increased competition. In Brooklyn, a new line introduced a six-cent fare in 1847. In Pittsburgh, fares remained at twelve cents into the 1850s but were probably cut after horsecar competition began in 1859. Given the average wages of a dollar a day or less for laborers and two dollars a day for craftsman, it is unlikely that they could regularly afford to spend twelve cents or even half that to ride the omnibus to commute to work. Most laborers undoubtedly walked to work, as they had always done. Members of the working class probably did occasionally ride the omnibus for a holiday outing, to avoid bad weather, or even for shopping, but regular patronage was unlikely. As one urban historian notes, the omnibus was "an improved version of the private vehicle, a better mode of conveyance for well-to-do people on flexible schedules."[25]

The omnibus provided a convenience for a segment of the public that could afford the fare but couldn't afford carriages. As time went on, this segment became larger because competition often drove fares down and the middle class was growing. The omnibus lines served the needs of those businessmen and professionals who wanted to separate work and residence. As early as 1837, one New Yorker wrote that omnibuses were "particularly convenient for merchants and others doing business in the lower part of the city, and living in the upper part."[26] In 1849, the suburban *Brooklyn Daily Eagle* observed that only "a few minutes' ride in the omnibus" put the passenger "in the business part of a large city."[27] Others who benefited included women shopping in the retail districts downtown, those needing to transfer messages or documents, and pedestrians seeking to avoid inclement weather. Still, the omnibus fare, whether twenty-five cents or twelve and a half cents, was too high to maximize the horse's centrifugal effect on urban shape. Reducing the number of horses per vehicles was probably the most obvious path to lower costs and fares.

The Coming of the Horsecar

While the omnibus was a substantial improvement over the limited urban transport that had previously existed, the development of the railed streetcar represented a great leap forward. The comfort of its ride was far superior to that of the omnibus, and it traveled at a more rapid speed. It was to have major spatial, social, cultural, and political consequences for American cities.

In 1831 the New York State Legislature granted the first streetcar charter in the United States to the New York & Harlem Railroad, permitting it to run cars on tracks on New York City streets. The New York & Harlem Railroad began operations in 1832, extending a track from Pine Street to Fourteenth Street; by 1839, it was operating passenger cars between City Hall and the village of Harlem. The railroad used steam locomotives from Harlem to Twenty-seventh Street, but south of there a city statute compelled it to substitute horses. In 1838 the railroad possessed one hundred horses, forty cars, and four locomotives, with a depot and stables at Fourth Avenue and Twenty-seventh Street.[28] John Stephenson, who had just built the city's first omnibus on Parisian lines, designed and constructed its first horse-drawn rail car. He modeled it after the English four-wheeled passenger railroad car but dropped the body down over the wheels for easier access. Four horses pulled the car, and it carried thirty passengers.[29]

The New York & Harlem Railroad ran its cars along iron tracks fastened to granite sleepers that rose several inches above the surface of the street. These tracks provided a smooth ride for the vehicle on the rails, especially compared to an omnibus running on the rough street. The *New York Daily Advertiser* enthusiastically wrote of the new line that the "horses appeared to fly with a load, which, if on the pavement, they could not have drawn."[30] The tracks, however, greatly hindered crosstown traffic, jarring the cabs, carriages, and wagons that attempted to cross them. Initially, horsecar drivers would pick up and discharge passengers wherever they wished, but later the road limited its stops to specified street locations, usually several blocks apart. The railroad's patronage greatly expanded, carrying both local passengers within the city and passengers commuting from the northern end of Manhattan to offices in lower Manhattan. The local route carried more than a million passengers by 1840 and 3.5 million by 1859. In the early 1850s fares were six cents for local passengers; cars came at five- or six-minute intervals during most of the day. This met the criteria for effective urban transport for those who could afford the fare.[31]

Steam locomotives on crowded city streets presented hazards for urbanites. Locomotives normally pulled cars as far south as the New York & Harlem's Twenty-seventh Street depot, where horses usually took over. Occasionally, however, a steam locomotive would draw a heavily loaded train below Twenty-seventh Street in Manhattan, but statutes forbade them south of Fourteenth Street. In 1839, the boiler of the locomotive New York exploded at Fourteenth Street and Fourth Avenue, in a well-to-do residential neighborhood, killing the engineer and injuring twenty people. Seeing their worst fears concerning steam-powered technology realized, neighbors reacted with fury at the deaths and injuries, threatening to destroy the tracks and the cars. Steam locomotives also occasionally killed or injured pedestrians who stumbled in their path, and New Yorkers increasingly questioned the wisdom of running them at street level. In 1844 the New York Common Council prohibited steam locomotives south of Thirty-second Street, a limit extended to Forty-second Street in 1856, although the road often violated the prohibition.[32]

In the 1840s New York entrepreneurs obtained permission from the Common Council to construct a horsecar line on Eighth Avenue. Omnibus owners complained bitterly that the railroad was stealing their passengers and for several years blocked proposals to run rails on more New York streets.[33] By 1850, however, the opposition had been overcome and the Common Council had granted franchises to the Second Avenue Railroad Company, the Third Avenue Railroad Company, and the Ninth Avenue Railroad Company. Fearing that the omnibus was doomed as a major carrier, some omnibus line owners joined with those proposing to construct rail lines on both the West and East Sides. Others sold their horses and equipment to their competitors, while those who retained their omnibuses cut their fares and focused their operations on uptown and crosstown streets without horsecar lines. In some cases the Common Council required new street railway companies to purchase the assets of the old omnibus companies.[34]

In 1857 the new horsecar lines carried more than 23 million Manhattanites, the great mass of the riding public. A horsecar could carry about three times as many passengers as an omnibus, provided a more comfortable ride, and was easier to enter and leave because it was nearer the ground. Horsecars charged only a nickel fare, less than half that of normal omnibuses fares.[35] By 1880, New York horsecar ridership had expanded to 160,952,832 passengers, with 11,760 horses and mules pulling 1,493 cars over 136 miles of track.[36]

Horsecars usually required fewer horses than the omnibus to haul an equal or larger number of passengers, and this need for fewer horses permitted a lower fare, making the horsecar available to a wider spectrum of the public. Part of the lower cost structure also resulted from better operating surfaces, including rails and superior pavements for the horse to trod upon. These improvements resulted in reductions in vehicle weight. Both eased the workload of the horses and allowed a reduction in their numbers.

John Stephenson, the New York coach and omnibus builder, had a virtual monopoly on independent horsecar construction before 1870. His first street railway car was nothing more than three coaches with side entries bolted together, running on iron wheels with flanges to fit the tracks. Thus, the vehicle was compartmentalized like the passenger cars on European railways, each compartment limited by fare to a specific class. Such heavy vehicles required four living machines. Between 1855 and 1865, according to data provided to the New York state engineer, New York City street railways reported a drop in the weight of twenty-passenger horsecars (almost all were Stephenson products) from 6,800 pounds to 3,500 pounds, nearly a 50 percent weight reduction. Street railways, which had originally operated with two or four horses per vehicle, now needed only two horses per vehicle. While both horses had to be very large, big animals were becoming increasingly available because of special breeding. Lower operating costs made street railways economically feasible on streets with lower traffic densities than Manhattan's north-south avenues. Stephenson-made cars operated on the initial street railway lines, not just in most North American cities but also in such distant spots as Caracas and Bombay.[37]

Stephenson lowered the weight of his cars through a series of steps: he replaced oak with hickory (or ash when possible) and later steel for weight-bearing parts like axles; he replaced wooden sides with tightly stretched canvas and added larger windows, glass being lighter than wood; and he replaced the compartments, which had multiple side doors, with a single rear entry. In some spots, he literally cut corners. He placed seats parallel to the sides of the vehicle so they did not need separate backs and curved the side of the vehicle under the seat. This also reduced the area needed for relatively heavy flooring. By placing wheels under the indented side, he shortened the wheelbase. Narrowing the wheelbase shortened axles, the heaviest component of streetcars. He also dropped the roof above the seats, decreasing the height of the sides. The roof remained raised over the central aisle, where passengers stood. This clerestory roof typically had lightweight glass sides. Beginning in 1859, he made open cars (with no sides at all) for the heavy summer business.[38]

In 1860 New Orleans entrepreneur Jack Slawson developed even lighter twelve-seat vehicles, the so-called bobtail cars, which weighed less than twelve hundred pounds. Their slang name derived from their appearance: stairs leading to their back doors looked like a bird's tail. The stairs replaced the customary (and heavy) rear platform. Bobtail cars allowed one-horse, one-man operation, an especial attraction in the South, where relatively light mules were the primary power source and ridership was not as heavy. In the North bobtail cars served lightly traveled suburban routes and late-night runs. By 1883 two-thirds of the cars that Stephenson was making were bobtail cars. In every city where the bobtails were introduced, however, the public protested, finding them more crowded and slower. But managers liked them because they could rest horses instead of running two-horse service late at night or on lightly traveled lines and because they reduced labor costs.[39] City leaders turned to these technological solutions, implemented by private entrepreneurs, to deal with population growth, increased density, and emerging industrialization. These innovations invariably clashed with older ways of doing things, as more traditional aspects of urban social life and customs, such as the use of the streets for social purposes, were threatened and transformed by the forces of technology and modernization.

For readers in the twenty-first century, it is perhaps hard to think of the relatively slow-moving horsecar as an instrument of change and modernization. Contemporaries, however, often saw it that way. Alexander Easton, a Philadelphia advocate of streetcar adoption in the late 1850s, for instance, emphasized the manner in which the horsecar operated as part of a functioning machine:

> Time is economized by regularity of transit; the car being quickly stopped by the application of the brake, the most refractory horses are immediately arrested; while the whole operation becomes so mechanical, that the horses, when accustomed to the signals of the bell, stop and start without any action on the part of the driver, by which means a timetable can be effectively used.[40]

Others emphasized the almost complete absence of friction between the wheels and the tracks and compared the ride to that of a sleigh "on ice."[41] A Baltimore commentator observed that a streetcar could carry three times as many passengers as the omnibus and was easier of access, "consequently you have one beautiful vehicle gliding gently along, instead of three cumbrous [*sic*] machines jolting you over rough pavement."[42]

On the eve of mechanization of street railways in 1887, the *Engineering News* noted the influence of the horse railway with amazement: "We are accustomed to think of 1860 as a tolerably civilized and advanced period. That there has been growth since that date is well understood, but that there has been a growth so vast that 1860 may be looked on almost as the beginning of civilization as respects internal city traffic, and all before that almost as a period of barbarism, is hardly realized."[43]

While it is difficult to accurately estimate the magnitude of the throngs pulled by horses on rails because of variances in census reporting and issues relating to new technologies, company mergers, and boundary changes in cities, one can provide an order of magnitude. In 1880, for instance, New York ridership per capita was 127, and by 1890 it was 297, or an increase of 134 percent; in Philadelphia, 1880 ridership per capita was 104, while in 1890 it was 158, or a 52 percent increase. And ridership per capita in Pittsburgh increased from 127 in 1880 to 193 in 1890, or 52 percent. The figures are a little distorted, but only slightly so, because 15 percent of the nation's street railways were mechanized by 1890. Almost all of this ridership change reflects the effect of the living machine—horses were pulling many more humans than they once had.[44]

Spread and Sprawl

Different cities experienced the effects of horsecars in various ways, depending on factors such as neighborhood opposition to the laying of tracks and the granting of franchises, political factionalism and political leadership, the availability of capital and entrepreneurial skills, population densities, and street patterns, topography, and geography. New York, Boston, Pittsburgh, and Philadelphia all had somewhat different patterns of streetcar development and construction.

The New York horsecar lines had the most influence on the development of areas of Manhattan above Forty-second Street. By opening up this territory, the new transit system kept those desirous of a suburban life on the island and still paying taxes to the municipality. In addition, space formerly devoted to residences in the crowded downtown was now freed for commercial development, and the downtown was prevented from "choking on its own congestion."[45]

Other cities rapidly emulated the New York horsecar lines. Boston entrepreneurs inaugurated that city's street railway era in 1853, when the legislature approved two lines, both displacing omnibus routes. Within a year the Cambridge Railroad Company was operating thirty cars pulled by 220 horses on cast iron rails and transporting an estimated five thousand passengers daily.[46] By 1860, fifty-seven miles of lines carried 13.7 million passengers.[47]

In 1850 Boston was a walking city hemmed in by tidal basins, marshes, and the ocean, with dense residential and commercial patterns. Workers' "barracks," tenements, workplaces, factories, and commercial buildings, along with middle-class homes, filled Boston's narrow streets. Citizens initially opposed the streetcars because of concern over crowding and risk on the constrained and winding streets. Before 1873, unlike New York, the Boston horse railroads did little to advance the city's settlement patterns. Their tracks followed old, established paths and by 1873 had expanded the periphery of dense settlement only from two to two and a half miles from City Hall. The greater effect of the horsecar came between 1873 and 1887, later than in New York, when better transportation extended the zone of new suburbs to four miles from City Hall.[48]

Crowding in the streets of the city's old core expanded greatly, driven by increases in population and numbers of vehicles. Because so many Boston streets were too narrow to accommodate flows of horses, carriages, wagons, and horsecars simultaneously, heated struggles between vehicles were common. But demand for transit was high, and by 1880, 551 horsecars carrying 26,953,540 passengers ran on Boston streets. The average number of passengers carried per car per trip was thirty-five.[49] In 1887, after years of fierce competition between firms, the entrepreneur and speculator Henry M. Whitney used his hold over the West End Street Railway to purchase control of five other Boston operating companies and to form a traction monopoly. Whitney believed that the way to profitability was to expand his ridership by extending his lines and retaining the five-cent fare.[50] Like the rest of the street railway industry, he firmly believed in charging a flat five-cent fare, regardless of the distance traveled. This policy represented a subsidy from inner-city, short-haul travelers to largely middle-class suburban commuters. It also served Whitney's other interest—speculation in suburban land—by making it accessible to downtown.

The streetcar arrived relatively late in Pittsburgh (1859), after the development of steam railroad connections with Philadelphia and other cities. As in other cities, complaints about "tardy and annoying" omnibuses encouraged the introduction of streetcars. The Pittsburgh horsecar lines followed essentially the same routes that had been used earlier by turnpikes and omnibuses—those of least travel resistance through the city's rugged topography. The City Council chartered four lines in 1859, all running from the downtown business district to outlying towns. Because of the hilly terrain, streetcars were smaller in Pittsburgh than elsewhere. Two horses or mules pulled most cars, although some lines running in flat areas utilized only one horse.[51]

By 1863, the Pittsburgh region had more than 16 miles of track. In that year, the first for which statistics are available, the four lines carried 3,960,009 passengers, an average of 40 rides per inhabitant. In 1871, the system had grown to 21.3 miles, with 50.6 rides per inhabitant; in 1880, rides were 61.7 per inhabitant with 11,885,980 total passengers; and in 1888, the last year of the horsecar era, the system consisted of 55.85 miles of track carrying 23,456,456 passengers, or 68.2 rides per inhabitant. The number of horses and mules used in the 1880s ranged between one thousand and seventeen hundred. As in other cities, ridership had increased between 1863 and 1888 at a much faster rate than had population, a sign of suburbanization. During the same period, the number of horsecars grew from 67 first-class cars (two-horse) and 12 second-class (one-horse) cars to 134 first-class and 38 second-class cars. Presumably the second-class cars were bobtails.

During the years when horses and mules provided the "motors" for streetcars, the transit companies extended the original four lines to a distance of five miles from the downtown, while other, newly chartered companies built crosstown and feeder lines. (The City Council approved charters for ten additional companies.) Fares varied from five to six cents, with two cents charged for transfers. Not all lines, however, offered transfers, a cause for much complaint by the riding public. The extension of horsecar lines facilitated the spread of population. Wards within three or four miles of the downtown core (an hour's horsecar ride) experienced heavy population growth, with extensive building along the streetcar routes within a five- to ten-minute walk of the tracks. The city's population grew from 49,601 in 1860 to 238,617 in 1890, much of it because of annexation of contiguous towns located on streetcar lines. Densities naturally thinned out farther from the radial rail lines, leaving broad underdeveloped areas distant from the core. In other words, Pittsburgh was both growing and deconcentrating.[52]

Horsecar lines were especially important in facilitating the movement of population from the densely congested core into the newly annexed areas, especially on the east side of the city (the flattest territory). In 1888, for instance, just before some lines adopted the cable while others electrified, the four lines running east on the Pittsburgh "peninsula" (i.e., the areas between the Monongahela and Allegheny rivers) carried 43.9 percent of the total passenger traffic. Housing construction boomed outside the core and especially in the recently annexed east end (wards 13–23), which saw an increase in the number of dwellings from 5,350 in

1870 to 17,604 in 1890. Simultaneously, the original core of the city lost residential population as it was transformed into a central business district (CBD). Contractors built many new buildings, a significant number of which were five- and six-story office structures in the emerging CBD. By the 1880s, the downtown and the outlying residential wards, connected by streetcar lines, experienced daily flows of primarily white-collar riders commuting to work and shoppers traveling to downtown stores.[53]

Examination of system maps of other cities reveals similar patterns. Horsecar lines always radiated out from the center of cities like the spokes of a wheel, except where natural obstacles like broad rivers or arms of the sea intervened. In Philadelphia, improved urban transportation created "bourgeois corridors" used equally by middle-class men and women.[54] The horsecar stimulated the development of these type of corridors in most sizable cities. In cities with hilly terrain, like Boston and Pittsburgh, horsepower limits required streetcar routes to follow the flattest available path, and suburban development concentrated in the valleys.[55]

Thus, many urbanites found that the horsecar improved their ability to navigate the urban environment. In February 1859, for instance, Sidney George Fisher wrote in his diary that the horsecars were a "great convenience," having almost "displaced the heavy, jolting, slow and uncomfortable omnibus." Fisher observed that the cars were "roomy, their motion smooth & easy, they are clean, well cushioned & handsome, low to the ground so that it is convenient to get in or out and are driven at a rapid pace." He also predicted that they would transform the character of urban living, since everybody would have a "suburban or villa or country home" and the benefit of "pure air, gardens and rural pleasures." Downtown would become a "mere collection of shops, warehouses, factories and places of business."[56]

Resistance to the Streetcar

The adoption of the horsecar provided many advantages to city dwellers, making the twenty- to thirty-year lag between New York's development of a horse-drawn streetcar line (the New York & Harlem) and its emulation by other cities or even elsewhere in New York City difficult to understand. Why the delay in technology transfer? Several factors seem most critical. One was that the New York & Harlem horses pulled not only heavy passenger coaches but also freight cars, obscuring the advantages of specialized, lighter horse-drawn passenger railways, although providing revenue for the road.[57] A second factor was the popular feeling against railroads on city streets because of concerns over risk and pollution. Third was the opposition of vested interests, especially the omnibus companies. While some chose to invest in street railways, many did not. An additional element may have been technical—that is, the problems caused by streetcar tracks, especially for crossing traffic. And the capital costs of the systems and projected ridership limits must also have been factors. Considering the amount of opposition to the horsecar, the delays are understandable. The authorities in London and Paris, for instance, never allowed horse railways downtown, preferring omnibuses.[58]

An 1866 survey of the horsecar experience in eleven American and Canadian cities, done by a British firm (John Noble & Company), provides some of the answers. The survey examined New York, Brooklyn, Boston, Philadelphia, Baltimore, Providence, Newark, Chicago, Quebec, Montreal, and Toronto, and it was done to find evidence to persuade Parliament to approve London streetcar lines. All these cities reported initial opposition by various stakeholders. Omnibus companies "opposed the scheme most vehemently," especially in Boston, New York, and Providence. Reports from Brooklyn and Quebec related opposition from storekeepers, who feared a loss of business from added traffic, while cabmen and carters in Montreal and Quebec feared both traffic and competition.[59]

The survey recorded diffuse popular opposition in several cities, especially Philadelphia.[60] In that city horsecars threatened street uses familiar to many residents. These customary uses of streets included "socializing," a commonplace in working-class neighborhoods; the formation of an architectural baseline, which dictated the arrangement of building sites, an important concern in elite neighborhoods; "promenading," or informal processions that defined membership in the bourgeoisie class; and hauling, especially in commercial and manufacturing districts.[61]

Opposition in Philadelphia continued an earlier pattern of antagonism to freight railroads and citizens' resentment of railroads as "corporate usurpers of what they regarded as common public space."[62] More specifically, opposition against the granting of charters to Philadelphia street railways was often based on the character of the specific streets on which the streetcars would run. For instance, in 1858 opponents to street railways on fashionable Chestnut and Walnut streets warned that their introduction would be "a barbarism" and would threaten shopping in "rich fancy stores," "promenading," and the operation of private carriages. One protest warned that streetcars would "invade, vandalize, and vulgarize our choicest streets or public spaces."[63] Petitions to the state legislature concerning other streets emphasized the interference with walking and threats to women and children patronizing the city markets. Still others warned that street railways would disrupt the normal traffic of drays, wagons, omnibuses, and coaches, as well as the loading and unloading of freight in a city with very narrow streets. More traffic would increase the risks associated with the living machine: biting, kicking, and running away.[64]

Ultimately, in Philadelphia as in other cities, these arguments failed to counter the political weight of the streetcar advocates, and the city councils granted the necessary franchises. In January 1859, the mayor applauded the City Council's actions and commented that "perhaps no public improvement ever occasioned more contrariety of opinion than the occupation by the passenger railway system of the streets of this city, and perhaps none has ever promised more general benefit to the community."[65] By the 1850s, the courts were defining streets "as having a single overriding function: the transportation of people and goods"—a new definition that drove out the older street uses.[66] The new values that the railroads and their users emphasized in determining street functions were those of "efficiency, speed, and progress."[67] These were the values that characterized the emerging networked city, girded by systems of rails, water supply pipes, sewers, and wires carrying street cars, water and sewage, and telegraph messages.[68] Older, preindustrial values, especially those regarding the uses of the streets for play, promenading, and walking, were increasingly threatened.

The efficiency of the railed system and the smoothness of the ride, however, could be exaggerated. First of all, while the proponents of the horsecar tried to present the horse as part of the streetcar machinery—a living machine—and deemphasize its animal nature, horses were still animals and resisted control. Not only did they occasionally run away, but they also polluted the street with their manure and urine and frequently died in the traces. (Of course, these were not only "animal" characteristics but were also characteristic of machines such as the automobile!) Another issue was clearly technological—the design and construction of the tracks and the traffic problems they created. This was a major issue in several cities. Philadelphian Sidney George Fisher, for instance, while praising the streetcars, also noted that they did "obstruct the streets for carriages. The rails make driving very inconvenient & unpleasant."[69] Here, then, lay another source of potential conflict—one between horsecars and private horse-drawn vehicles, such as carriages and wagons. Ideally, in the construction of a smooth iron guideway for horsecars, the rails should interfere minimally, and possibly even facilitate, other street traffic. Unfortunately, the opposite was often true.[70]

Tracks varied in type but essentially involved iron rails or strips of different designs fastened with spikes to longitudinal wooden beams called stringers. The stringers then were placed across wooden crossties about three feet apart and laid in a ballast of broken stone or coarse, sifted gravel. Between the rails, companies primarily used granite blocks, a paving innovation of the 1850s that provided horses with good footing. Pavements covered the foundation and crossties. Most wheels were flanged, and these flanges ran in grooves in the rail, but the depth of the grooves differed considerably. City ordinances normally set track gauges, as well as other standards relating to tracks and the street. Track gauge was variable, ranging in width from four feet six inches to five feet two and a half inches, with many tracks at four feet eight and a half inches, the standard steam railway gauge. In some cities, such as Boston, track gauge actually varied from line to line.[71]

Other horse-drawn vehicles found the rails both a problem and a convenience. When they could, carriages and wagons tried to run on the tracks to benefit from the smooth surface, but in so doing they often interfered with streetcar operations. In 1855, the *Brooklyn Daily Eagle* commented on "the daily practice of carriages and private teams to take possession of the track and purposely keep in front at such a snail-like pace as to seriously annoy the drivers and incommode the passengers in the cars."[72] In 1881, an observer from the North Chicago City Railway counted 2,052 vehicles running on the tracks on a busy street over a ten-hour period, only 642 of which were streetcars. This suggested to the company engineer that street railroad tracks needed to be designed to carry all types of traffic.[73]

This use of the rails caused numerous problems when drivers of drays and carriages failed to get all four wheels on the tracks, running one set of wheels on one track and the other on the surface. The friction of the wheel on the pavement dug a narrow and rough trench that vehicles had problems exiting. An 1859 *New York Times* editorial noted that these troughs (one and three-quarters to three inches in depth) obstructed and strained all the wheels that crossed them. The paper advised substitution of a "wide-flat" rail five or six inches wide that only rose three-quarters of an inch above the street and had a groove for the flanges.[74]

Disputes over New York City traffic problems caused by streetcar rails flared in the press several times in 1860. "Tubal Cain," a knowledgeable *Times* writer, commented that the system of city railways represented a "compromise between the convenience of the largest number of passengers and the inconvenience of private and goods traffic." The tracks, he said, caused havoc for cross and diagonal traffic by catching and damaging the wheels of vehicles that tried to turn, breaking axles and creating a "broken and gullied pavement." These rails, said Cain, "were adapted to carrying flanged wheels, and to rapid destruction of all others."[75]

Cain pointed to Philadelphia (ironically, considering the complaints of Sidney George Fisher) as a city where streetcar rails did not create traffic problems because they were configured to accommodate both streetcars and other traffic. In 1855, Strickland Kneass, Philadelphia's chief engineer and surveyor, had recommended to the City Council that it require street railway companies to use a rail that was low enough to permit "ordinary vehicles to cross and recross." He suggested a track gauge that conformed to those of ordinary carriages. In this manner, he noted, "vehicles traveling upon the track, will use both rails without injury to the iron or pavement, and the wagon-edge of the rail will offer no impediment to turning out."[76] Philadelphia streetcar companies followed Kneass's recommendations, laying tracks that were five feet two inches wide and provided a tramway that ordinary vehicles besides streetcars could use.[77]

But rail designs varied widely from city to city until the 1880s, when many companies substituted steel for iron rails and adopted the Johnson girder rail. This rail, invented by Tom Johnson of Cleveland, did not obstruct traffic and eliminated the need for stringers.[78] In more than one hundred small cities, however, companies laid tracks on unpaved streets and installed the obstructive "Tee rail," often leading to serious problems.[79] As street railways became more common in American cities, the problems they created rather than their advantages became more obvious. Sidney George Fisher, who had praised the streetcars when they were first introduced, now criticized them because they ruined "the road . . . for driving" and because they subjected government to a "system of bribery & corruption" in their quest for franchises. Like others confronted by the advantages of modern technologies but also by the possibilities for evil by the corporations that controlled them, he was torn between the alternatives. By 1866 he despaired that "the companies do as they please & the people submit to high fares, crowded cars & almost impassable & often dangerous streets & roads."[80] What he had originally viewed as a marvelous innovation had now become a force for corruption and a threat to the safety of the streets.

Why Fisher saw the streetcar companies as forces for corruption requires an understanding of the broader compass of American urban history during the second half of the nineteenth century. Growing populations and expanding retail, commercial, and industrial sectors demanded new infrastructure, such as water and sewer systems, gas lines, paved streets, and fire and police alarm telegraphs, as well as street railways. Since control over the street, the site for most infrastructure whether above or below ground, was in the hands of the municipality, many potential opportunities for graft developed in the awarding of franchises and contracts to corporations seeking monopolistic control.[81]

Cities with well-developed governmental institutions and strong leadership might have been able to resist the descent into the buying and selling of government privileges, but growing American cities were in the process of shifting from a form of urban governance with primarily patrician leadership to one dominated by the political machine catering to a large and heterogeneous public electorate. In this context, legal procedures were frequently violated and votes sold to the highest bidder. One of the most lucrative sectors for machine politicians related to the use of the streets for infrastructure. Many urban politicians became political entrepreneurs, using their votes as investment capital and selling them to the highest bidders. The political machines that emerged systematized rather than eliminated the buying of votes and thus acquired funds to perpetuate their political power with the voters. Potential franchisees could also give jobs to followers of their political supporters, a somewhat less reprehensible practice.[82] Street railway franchises thus became a prime commodity on the governmental market early in their history.

Who Did the Horses Pull?

Increases in ridership lead to the question of who rode the streetcars and why? The horsecar clearly transformed the meaning of urban space and advanced the process by which the compact walking city became the extended, networked city. The extent to which the horsecar lines contributed to suburbanization, however, is difficult to estimate. Those commuting daily between residence and workplace must have accounted for a significant number of riders. Clearly, many other passengers, especially women, utilized the streetcar to journey from their homes to shop in the growing downtown districts with their new department stores.[83] How many of these passengers had opted to move farther from the core of the city toward its outskirts because transportation had become available, however, has not been systematically evaluated.

The separation of residence and workplace was a major development of the second half of the nineteenth century. Owners and managers of firms and bankers and professionals such as lawyers (but not physicians) increasingly separated workplace and residence, moving to the fringe of the city and to the expanding suburbs. The growth of the factory system and of large industrial workforces also resulted in further separation of workplace and residence for many employees. In some cities, such as Chicago, Pittsburgh, and Toronto, workers also moved to suburbs, either following factories that had moved to the fringe or because they were attracted to the possibility of home ownership.[84]

It is possible to trace the journey-to-work (JTW) pattern for professionals and some businessmen who have identified workplaces (with residences available from the manuscript census), but the task is much more difficult for members of the working class. A basic element for detailing the JTW—place of employment—is not available for most horsecar riders. Which groups in the urban population utilized the streetcars is an even more difficult question, and while there is considerable anecdotal data considering the character of the riding public, quantitative data are not available.

One important study of Philadelphia, however, estimates that, from 1850 to 1880, the JTW doubled in length for those who labored outside their homes. The authors explain that the horsecar lines were only one factor in the lengthening of the JTW. Other factors were a post–Civil War housing construction boom, which led to the creation of new neighborhoods; the growth of a downtown or central business district, with many nonresidential land uses and a declining resident population; the growth of an industrial economy with large factories; and a faster spatial expansion of population than of industrial jobs. These factors combined, they explain, "to leave people further from their places of employment in 1880 than in 1850."[85]

JTW distances differed, however, for various professionals—physicians largely combined work and residence in both years, but lawyers increasingly separated the two. Public servants and bank employees expanded their JTW, while the patterns for small proprietors and artisans depended on the extent to which they served local markets. Blacksmiths, for instance, remained in all the neighborhoods examined, reflecting the continued importance of horse-drawn transportation. The JTW for the growing industrial workforce (based on a sample of four thousand workers) approximately doubled during these years—from a half mile or less to about a mile, although industrial workers largely continued to walk. The average unskilled, semiskilled, and even skilled worker simply could not afford the streetcar fare on a daily basis.[86] In addition, the layout of Philadelphia streetcar lines did not lead to efficient connections between residences and industrial workplaces. Riders often had to ride more than one line to reach their destinations, and many companies refused to provide transfers for competing lines, raising the cost of the trip.

The Philadelphia study concludes that only 17 percent of employed persons or 7 percent of the total population would have regularly ridden streetcars for their JTW. These commuters were largely nonmanual workers, living over a mile from their workplaces. The largest percentage of passengers rode the horsecars for downtown shopping trips (28% of riders), followed by those accessing recreational facilities (25% of riders, all days of the week) and intercity travelers moving between railroad stations and ferries (10% of riders).[87]

Transportation innovations in Philadelphia, however, "led the way" in driving new land use patterns and expansion of the city's residential, industrial, and commercial sections. A housing boom outdistanced population growth, leading to reduced residential densities and growth at the city's fringe in suburban environments. Other studies also confirm that the movement of higher-income groups in Philadelphia paralleled the direction of the streetcar lines, as affluent residential populations moved out of the core of the city and toward the northwest.[88] The central business district (CBD) lost residential population and gained commercial activities, including retail, wholesale, and banking firms, while white-collar workers employed in the growing CBD increasingly commuted to work.[89]

Studies of New York City reveal a similar pattern. The omnibus, the steam railway, and the ferry had already made possible some residential dispersal, and the horsecars accelerated the process. An 1855 article in *Putnam's Magazine* estimated that seventy thousand people commuted to the city by those travel modes. By 1865, only 2 of the city's 1,348 stockbrokers lived in ward 1, the locale of Wall Street. While lower-income workers still lived near their places of employment on the Lower East Side and what is now called the East Village, they were increasingly distant from upper- and even middle-class neighborhoods.[90]

Other studies of Milwaukee, Pittsburgh, and Toronto generally confirm the pattern described for New York and Philadelphia. That is, these three cities in the horsecar period (1860–90) experienced the growth of outlying residential areas, the development of streetcar lines through the city and toward the urban fringe, an increased concentration of retail, financial, and corporate interests in the urban core, and a decline of downtown residential population. Like Philadelphia, the lines were used for recreational, shopping, and commuting purposes, although the share of total ridership for each function is unknown.

The horsecar alone did not create these changes in the urban fabric—other factors helped drive them, notably economic growth, population expansion, and a desire for suburban homes located away from the dense and polluted streets of the walking city.[91] The horsecar, however—cheaper, faster, and more comfortable than the omnibus—was an important facilitator of middle- and business-class suburbanization. Those involved in suburban real estate development grasped its possibilities, and newspapers were filled with ads detailing the location of suburban homes within easy reach of horsecar lines. In many cases, the owners of the transit lines were also owners of extensive suburban real estate. Transit lines, therefore, "led rather than followed subdivision . . . [and] were tools in a competitive real estate market."[92] Thus, as one observer commented, "It is hardly too much to say that the modern horse-car is among the most indispensable conditions of metropolitan growth. In these days of fashionable *effeminacy* and *flabby feebleness,* which never walks when it can possibly ride, the horsecar virtually fixes the ultimate limits of suburban growth."[93]

As urban population spread, industry moved farther from the CBD and employed larger numbers of workers per firm. In Pittsburgh, the growing iron and steel industry led this trend. The great majority of laborers in these industries almost certainly walked to work, a pattern continuing well into the twentieth century. Toronto workingmen appeared to follow a slightly different pattern, with working-class members who had moved to the urban fringe using the street railways for their JTW. In 1891, for instance, when the municipality bought the streetcar lines from the private companies, it also provided for workingmen's fares. Such arrangements were uncommon in the United States.[94]

Social Conflict

Equine power sources allowed the residential segregation and private homes so avidly sought by a middle class beset by the rapid changes industrialization and migration had brought to cities. Paradoxically, however, horsecars and omnibuses threw together a variety of riders from different backgrounds in a common and confined space at exactly the same time that it was creating bourgeois utopias in the suburbs. As a result, vehicles frequently became a microcosm of the various forms of social interaction and conflict that occurred in cities.[95]

Some of this interaction reflected a simple lack of mutual understanding of acceptable public behavior among riders, acutely observed by the author William Dean Howells. Writing about horsecar trips from his suburban home in Cambridge into the city of Boston, Howells commented that the ride reduced the passengers equally "to the same level of melancholy" where "the courtesies of life are impossible, and the inherent dignity of the person is denied." Howells complained about crowding and occasional rudeness, but he found it particularly offensive that men refused to surrender their seats to women. For him the packed horsecar was a "prodigy whose likeness is absolutely unknown elsewhere." He predicted that it would be celebrated in the future as a "stupendous spectacle of human endurance," testifying to "the mystery of our strength as a nation and our weakness as a public."[96]

As Howells noted, the crowded horsecar did have a leveling effect, subjecting persons from different classes, ethnic groups, and backgrounds to a common but often friction-producing experience in a contested public space. In this way it resembled the city street at even higher levels of density. Unlike the city street, however, as Clifton Hood has observed, public transit riders "submitted to close bodily contact and class mixing in a space that was physically enclosed and socially contested along ethnic, racial, class, and gender lines." In 1874, for instance, the *New York Times* noted that some New Yorkers preferred to "walk any distance rather than endure the disgust and actual humiliation which even a short ride in a crowded street car entails."[97]

Women particularly experienced humiliation and embarrassment in the crowded space of the cars, especially during rush hours. Not only were they denied seats, but they were "pushed and crowded by men," subjected to "assaults" by gangs of drunk loafers, often touched inappropriately by conductors getting in or out of cars, groped by "bustle pinchers," and "have said to them the things that are uttered in this world only in an undertone." Women traveling alone "after dark," warned the *Brooklyn Eagle*, did so at their "absolute peril," as the streetcars became a "school of immorality."[98]

African Americans of both sexes, however, undoubtedly experienced much greater rejection and humiliation than other urbanites. Their riding experience reflected the deep urban divisions and tensions regarding blacks in the city. These conflicts are striking because they occurred before, during, and after the Civil War in the North and the South alike.

In New York City and Philadelphia, even before the streetcar appeared, African Americans had not been permitted to ride on omnibuses. They were also often forbidden to ride on New York streetcars, although the policy varied from company to company.[99] For instance, on a Sunday in the summer of 1854, a Third Avenue streetcar conductor forcibly expelled from his vehicle a black public school teacher on her way to church. She sued the Railway Company in the New York State Supreme Court for damages, the case being heard in February 1855. In his instructions to the jury, the judge held that the company was a common carrier and "as such bound to carry all respectable persons; that colored persons, if sober, well-behaved, and free from disease, had the same rights as others."[100] Thus, it appeared that African Americans who met these criteria would be permitted to ride on New York streetcars.

Still, that same year the Sixth Avenue Company began a new practice in regard to African American riders—the running of segregated cars. The company painted the notice, "Colored Persons allowed in this Car," on the side of three of their vehicles. The company's secretary claimed that this action was taken at the request of the trustees of the Colored Half-Orphan Asylum, "the more respectable portion of the colored people."[101] The streetcar company also permitted African Americans to ride on the front platform of any of its cars, while whites could also ride in cars designated for "colored persons."[102]

Sometime in the spring of 1855, a prominent African American Congregationalist minister, Dr. James W. C. Pennington, who was also head of the newly formed Legal Rights Association, delivered a sermon at Manhattan's Shiloh Presbyterian Church maintaining that blacks should assert their rights to ride in "all public conveyances." Several of his followers tried to assert this right, leading to their expulsion from streetcars and ensuing court cases. On May 24, 1855, Pennington himself challenged the company's segregation policy by attempting to board a Sixth Avenue car not designated for "colored" persons. The conductor ordered Pennington off the car and, when he refused, forcibly ejected him. Pennington attempted to secure redress from a policeman, but with no success, and launched a civil suit to claim his rights.[103]

In a letter to the *New York Times,* the secretary of the Sixth Avenue Company argued for a separate but equal doctrine. While the firm was a common carrier and had to provide equal "conveyance" to African Americans, it had the right to assign them to special cars. In defense of this policy, he argued that many riders objected to sitting next to a "colored man" or "allowing the ladies of their families to mingle with them in public conveyances." The company's business, he argued, was to "carry passengers, and . . . we have nothing to do with the color of their skins, only with the color of their money, and the comfort and convenience of all."[104]

Pennington responded in a letter to the *Times,* maintaining that the company's statement of policy concerning the irrelevancy of "the color of their skins . . . [was] the principle I contend for." He insisted that the company respect the law by not discriminating.[105] The company, however, refused to eliminate its segregated cars. The Superior Court heard the case of James W. Pennington vs. the Sixth-avenue Railroad Company on December 18, 1856. Here the defendant's attorney again maintained that the "great prejudice . . . among many people, about riding with colored persons" justified the policy of providing separate cars for blacks and whites.[106] The judge instructed the jury that, if it found that the "admitting of colored persons into the cars injured the business of the company in any way," they must find for the defendants. The jury so found, thereby having the effect, said the *Times,* of "prohibiting colored persons from riding in any of our public conveyances."[107]

The practice of providing segregated cars persisted in New York City into the

Civil War years on the lines of the Sixth Avenue and Eighth Avenue companies. An important 1864 case, however, involving an Ellen Anderson, the widow of a black Union Army sergeant who had been killed in action, affirmed the rights of African Americans to ride in unsegregated cars. In this case the conductor had ejected the plaintiff from an Eighth Avenue car (not marked for "colored" riders) with the help of a policemen. A company director denied on the stand that "colored citizens" were excluded from his company's cars, but the conductor maintained that, because there were special cars for "colored people," he had presumed it was a company rule that blacks could not ride in cars with whites.[108] The case was heard by the Police Commissioners, who affirmed that there "was no law against these people riding in the cars, and they had no right to make such a law." At the conclusion of the trial, the Eighth Avenue Railroad superintendent announced that "hereafter colored people are allowed to ride in all the cars, both small and large, of this company," ending the company's attempt at segregation.[109] In New York City, no further incidents involving attempts to limit blacks to segregated streetcars seem to have occurred.[110]

New York's experience in regard to streetcar segregation was relatively mild compared to Philadelphia's. Considered the most racist northern city, the city of brotherly love experienced fierce conflicts and violence over black ridership. When horsecars first appeared in Philadelphia in the late 1850s, eleven of the city's nineteen lines excluded African Americans, while the other eight permitted them only to stand on the exposed front platform. Two companies briefly attempted, like New York's Sixth Avenue line, to run separate cars for blacks. Challenges to these policies by African American spokespeople and their Radical Republican allies occurred on the cars themselves, in mass rallies, and in the courts throughout the Civil War. Violence, frequently directed against black soldiers in uniform who were attempting to ride the cars, often met these challenges. Adverse publicity in the local press and court decisions returning damages against the companies did not change their policies. Finally, in March 1867, faced by the prospect of black male suffrage, the Pennsylvania state legislature banned discrimination on Pennsylvania streetcars.[111]

In the southern states, patterns were somewhat different. Before the Civil War, in those few cities with streetcars, white owners often rode with personal slaves, although some cities, such as New Orleans, had separate cars for blacks.[112] In the postwar period, as Jim Crow patterns emerged, streetcars often became "racial battlegrounds," as blacks challenged attempts to relegate them to an inferior status.[113]

While streetcars in northern and western cities also experienced racial tension, especially during times of racial disorder, most conflict between the public and the streetcar corporations reflected issues of overcrowding, fares, and route expansion. Many cities also witnessed fierce labor conflict frequently involving the public. While of major significance to urban life, this area of friction and dispute has been dealt with in detail in other studies and will not be considered here.[114] Overcrowding, on the other hand, was endemic, especially during rush hours, in bad weather, and on resort routes on holidays, causing misery for horses and passengers alike. Riders often complained about the crush on the horsecars, while overloaded living machines found their advocates among animal rights advocates. Henry Bergh, the famous New York animal-rights reformer, once stopped a two-horse car with more than one hundred passengers on a snowy day. In 1875, in Providence, animal rights inspectors found two horses pulling 135 passengers, some clinging to the horsecar's roof, to a baseball game. These were only two of the many such incidents that occurred as more power output was demanded of the living machine than it could provide.[115]

Horse-powered transit had multiple effects on the city. As a vehicle that shaped space, it accelerated the sorting process of urban territory into specialized residential, commercial, industrial, and retail districts, stimulating the rise of a true urban downtown and allowing the separation of work and residence. To some, it appeared old fashioned because of its reliance on the horse, but as a railed technology it was a critical element in the rise of the networked city, facilitating and stimulating the development of other urban networks, such as water and sewer systems and the telegraph and telephone.

From the perspective of city residents, the horsecar brought both benefits and costs, thus reflecting the actual nature of urban residence itself. As a common carrier, it focused and exacerbated the tensions latent to the city involving class, gender, ethnicity, and race. African Americans probably experienced this tension more than any other group. For some, however, largely the middle class and elites, it provided mostly benefits: the ability to separate work and residence and to have access to suburban homes, better access to recreational facilities (see chapter 4), and access to downtown work, shopping, and entertainment. For members of the working class, there were also benefits, although limited by costs. Members of the working class undoubtedly used the cars in much the same way as did the middle class—to access shopping and recreation, especially on weekends—although on a much more limited basis. More highly paid working people used the cars to access work or perhaps occasionally to avoid exposure to foul weather.

Thus, as one attempts to estimate the effects of the horsecar on the city, it should be viewed as a path-breaking technology, setting in motion far-reaching changes in urban spatial structure. The horse retained its animal nature, but in its relationship to the streetcar it had become a machine and a critical source of transforming power.

CHAPTER FOUR

The Horse and Leisure

Serving the Needs of Different Urban Social Groups

The urban horse shaped the city in a spatial and economic manner, but it also filled a variety of leisure and recreational roles. These activities reflected the interests of different groups in urban society, each often possessing separate social and cultural meanings. Best known is the importance of the horse to elite leisure life and to the status-conscious who had a "frenzy for driving,"[1] but the horse also played a role in the leisure life of the middle and working classes. The urban horse, especially as the "motor" for the horsecar, made it possible for large numbers of urbanites to access various places of amusement distant from their homes. In addition, again through the agency of the horsecar, the urban horse enhanced the ability of members of the middle class to move to the suburbs and to embrace a suburban lifestyle. In this chapter we explore how the horse served the recreational and leisure lives of these different social groups.

Elite Carriage Driving

The great nineteenth-century economist Thorstein Veblen aptly summarized the relationship of the horse to the leisure life of the elite. "The fast horse," he wrote, "is on the whole expensive, or wasteful and useless. . . . What productive use he may possess, in the way of enhancing the well-being of the community or making life easier, takes the form of exhibitions of force and felicity of motion that gratify the popular aesthetic ends." Veblen noted that such display was an imitation of the British upper classes, commonly emulated by American elites. Horses also provided a spectacle—entertainment, if you will—for the non-horse-owning classes, who might admire the moving horse for "popular aesthetic ends."[2] According to a scholar of animal-human relations, these horse spectacles were a eulogy for the countryside and nature, serving a kind of rural nostalgia not only for their owners but also for those (perhaps recently arrived in the city from the country) viewing the spectacle of the elite parading their horses.[3]

Besides satisfying aesthetic needs, the horse had other leisure uses. The noted ecologist Paul Shepard has claimed that horses are inherently sensual objects because of their sleek coats and body curves and because of the genital stimulation experienced when riding. Another well-known student of animal-human relations notes that pets (and we would add, by way of extension, many horses) are "living art objects." This imagery is fully reflected in paintings and photographs that present scenes from another very important leisure application of the horse, horse racing. Racing in several forms became the first professionalized American sport and a tolerated form of gambling in many states.[4]

While the elite often used horses as status or prestige items, horses could also serve as living machines. Carriages themselves can be viewed as mechanical extensions of horses, along with such new attachments as padded horseshoes, calks, and blinders. Just as vehicle weight reduction and increased horse size facilitated horsecar development, they also allowed greater use of carriages. By 1870, but not before 1850, one or two horses could pull lightweight vehicles with seats for six and perhaps a driver and footmen as well. Better springs reduced the work of horses and facilitated a more comfortable ride, while glass windows made possible closed carriages. Just as wagons and horsecars morphed into a variety of shapes, so did carriages. In his 1867 book about carriages, Ezra Stratton, the editor of *N.Y. Coachmaker's Magazine,* listed thirty-one different types, only seven of which had existed before 1850. In 1910 Studebaker, the largest firm, made 115 different types of buggies as well as many types of carriages; there was one form of carriage for every taste, one for every purpose. Both the breed of horse and the style of carriage conferred status.[5]

The ostentatious display of horses and carriages was well demonstrated in the first years of New York's Central Park (created in 1857) when, as Roy Rosenzweig and Elizabeth Blackmar observe, "wealthy New Yorkers defined the new public park as their own." Designed by Frederick Law Olmsted and Calvert Vaux as a landscaped public space in the English tradition, in its early years Central Park was largely inaccessible by public transport. One of the initial Central Park guidebooks (1860) provided more information on how carriage owners and horses could get to the park than about access by public transportation. Olmsted prohibited omnibuses and express wagons from entering the park, as well as "any cart, dray, wagon, truck, or other vehicles carrying goods, merchandise, manure, soil, or other article," reserving its roads for those who could afford stylish horses. In its opening years conflict developed between Central Park gatekeepers and what the *New York Times* called "our wagon-owning citizens," who were aggrieved that they could not take a drive in the park on Sunday with "the respectable vehicles which do duty during the week in transporting legs of mutton, or cans of milk, or kegs of crackers, or boxes of candles from their shops to the customers' houses."[6]

During its first decade (1860–70), a substantial majority of the park's regular visitors arrived by carriage or horse to take advantage of the nine or so miles of carriage drive and the park's bridle paths. Few members of the middle class, much less the working class, could afford the expense of horses and the attendant carriages, stables, harnesses, and coachmen. Central Park, observe Rosenzweig and Blackmar, was "both effect and cause of a growing enthusiasm for carriages among the upper classes," as it stimulated a "revolution" in New York society. "Owning a carriage and riding in the park" identified one as a member of the city's upper class. Entrance into the carriage elite, however, rested entirely on the possession of money, since Central Park's carriage drives had no restrictions based on family background, origins of wealth, or religion. This new social world was dominated and shaped by women who served as "emblems of their husbands' wealth and judges of their own and others' status."[7]

Manhattan's old monied elite favored sedate black broughams pulled by huge horses. More chic individuals rode in vehicles in the latest Parisian fashion. Sporting types had light carriages pulled by trotters suitable to the race track. Jim Fisk, the shady but flamboyant Wall Street millionaire, had an elaborate equipage pulled by twelve horses—three pairs of black and three of white—with uniformed postillions in front and two footmen in the rear, probably the fanciest rig in town.[8]

In 1868 the autonomous City of Brooklyn finished Prospect Park (also designed by Olmsted) in an area then near the edge of the city but later destined to become its center. Olmsted added to his usual park drives by building a new transportation venue: the parkway. He added two of these roads, limited to carriages, to the park. The longer and more popular two-hundred-foot-wide Ocean Parkway ran straight for six miles from Prospect Park to the resort at Coney Island, opened in 1874. Shortly thereafter the *Brooklyn Daily Eagle* wrote: "The people who crowd Brooklyn's great drive daily are composed of gentlemen of leisure with their blood stock; men of business who leave their stores and offices after the heat of the day and in very many cases take their families with them for the benefit of a ride; men of limited means who spend all of their limited cash and sometimes more on horse flesh and the men who make their living training and driving horses." The article specifically noted that the new parkway, a "Paradise for Horsemen," was specifically designed for those who wished to surpass the rate of speed allowed in Prospect Park.[9]

The acme of the fashionable display of carriages was the coaching craze of the late nineteenth century. This involved driving a four-horse coach of the kind made obsolete by railroads, especially on a long intercity run. The pastime was extraordinarily expensive. Bostonian James Garland estimated the bottom line for keeping a coach and four at twenty-seven thousand dollars a year, well beyond the range of the merely affluent. Stabling on expensive urban land and feeding four horses was only part of the cost. Most participants owned more than four horses to allow relays on long runs. A full-time coachman cost twenty-seven hundred dollars a year and a hostler somewhat less. Not many people had the time to learn four-in-hand driving and join lengthy runs. For participants, this was part of its appeal—it required the ostentatious consumption of both time and money.[10]

T. Lawrence Bigelow of Boston had become the first amateur coachman in the United States in 1860, but the real growth of the spectacle came after several members of the elite Knickerbocker Club in New York founded the Coaching Club in 1875. Members included Manhattan socialites with such famous names as Belmont, Cassatt, Gerry, Havemeyer, Roosevelt, Schermerhorn, Tiffany, Vanderbilt, Whitney, and Woodward. All spent in excess of fifteen thousand dollars a year on their pastime, and each owned five or more carriages.[11] Thousands often gathered to watch the club's parades up Fifth Avenue and through Central Park. Its 1878 tour to Philadelphia was the first of many annual distance runs. Great care was taken to keep the schedule of the original intercity coaches, even to the point of sending servants out in advance with relays of horses. New York's sidewalks were packed for the departure of the wealthy owner/drivers. In Philadelphia, spectators "fortified by invigorating beverages" greeted the coaches with enthusiasm compared to that of the Fourth of July crowds. Evidently, many onlookers were more interested in seeing the finery of the women riding in the carriages, probably including the stylistically mandatory beaver hats, then the carriages themselves.[12]

The novels of William Dean Howells about Boston, Theodore Dreiser about Chicago, and Edith Wharton about New York provide clear documentation of horses and carriages as markers of status. Howells's *Rise of Silas Lapham,* originally published in 1885, is the earliest of these works. Howells's protagonist is Silas Lapham, a wealthy, self-made paint manufacturer, out of place among Boston's established Brahmin wealth. In the novel Lapham's horses and carriages play an important role, allowing him to overcome his plebeian roots by using personal transport on a level equal to that of the Boston Brahmins. Because of his roots in rural Vermont (he proudly proclaims that he was once a hostler and "always did love a good horse"),[13] he viewed fine horses as a form of status and consumption. He proudly showed his horses to a reporter writing a feature about his business, bragging that they were "Hamiltonians, with a touch of Morgan," both famous sires. He owned at least six vehicles, each for display on a different sort of occasion.[14] Interestingly, carriages seem to be one of the last things that men who experienced business failures were willing to give up.[15]

Howells has much fun throughout the novel satirizing the pretensions of the newly wealthy who flaunted their horses and carriages to assert their wealth and new elite status. Such display, however, must be viewed as a form of democracy, too, since anyone who could afford or even rent a carriage could participate in the park parade. In New York, for instance, figures of the city's underworld—like Madame Russell, a famous abortionist, and Josie Woods, the proprietor of Manhattan's largest brothel—were frequently seen in Central Park riding in their open carriages.[16] To be proper, however, carriages had to have the right style and decoration. Fashions in etiquette, in breed of horses, and in proper attire for coachmen and owners riding or driving changed constantly, not just from year to year but from season to season. George Washington's carriage, for example, could not be sold in Philadelphia in 1798 because of its ornamentation (green sides and red wheels). Republican simplicity demanded a less garish design. Except for a brief period in the 1860s, subdued colors were in fashion in the Northeast. In contrast, southerners supposedly preferred "the older, more flamboyant ways."[17]

The cult of the carriage also permeated Theodore Dreiser's 1900 novel, *Sister Carrie*. The novel traces the climb up the economic and social ladder of Carrie Meeber, a young woman from small-town Wisconsin, who moves to Chicago, becomes the mistress of a small businessman named Hurstwood, and rises to become a famous Broadway actress. On the evening their relationship is consummated, Hurstwood impresses Meeber by taking her for a carriage ride, carefully renting a closed carriage (i.e., one with sides and a roof, so the occupants could not be seen) at a livery stable. But after his conquest, Hurstwood returns to his wife and Meeber moves to New York, where she finds that the symbols of prestige are frequently horse related. She describes the shopping areas on Broadway "full of coaches [while] . . . Coachmen in tan boots, white tights, and blue jackets waited obsequiously for the mistresses of carriages who were shopping inside." She learns to measure status by the ownership of horses and carriages, as is reflected in the notes she receives backstage from "gentlemen with fortunes [who] did not hesitate to note, as addition to their own amiable collection of virtues, that they had their horses and carriages." At the novel's conclusion, her pursuit of wealth is fulfilled: "She could look about her gowns and carriage, her furniture and bank account."[18]

Edith Wharton presented a more nuanced view of the elite display of horses, in part because she was sensitive to matters of gender. In *The House of Mirth,* a high-society novel set in 1872, women travel only in open carriages ("an interminable procession of fastidiously decorated carriages") that allow them to display themselves to the public without actually going on the streets. One newspaper report on Prospect Park in the early 1870s observed an "increasing number of lady drivers" who drove pony phaetons. *Outing* observed the late-afternoon carriage promenade in Central Park: "All that is loveliest in womenkind, all that men envy most in their fellow man, all that is best in horseflesh is represented."[19]

Wharton also reflected her understanding of the importance of horses to status with her treatment of the annual horse show, which a wealthy group of New Yorkers began in 1883. The horse show came to mark the unofficial opening of the New York social season, and its directors and members formed the basis for Louis Keller's first *New York Social Register* in 1887. In *The Age of Innocence,* Wharton describes an exceedingly wealthy man as "the captor of blue ribbons at horse shows, . . . the owner of winning race horses," clear markers of status.[20]

Traditionally, riding horseback was beyond the pale of respectability for society women, although by the 1890s there was grudging acceptance for riding sidesaddle.[21] As late as 1895 one publicity-seeking Broadway actress contrived to have herself arrested for riding astride in Central Park.[22] *Rider and Driver* published articles that year urging the superiority of riding over exercising in a gym for women, since riders got more fresh air. The magazine even claimed that "the girl who can ride horseback is the one who gets the fellows."[23]

Not all observers, however, were fond of the spectacle of women controlling horses, and men often stereotyped women drivers as dangerous or nervous. One observer of park traffic noted that American women were poor drivers, with "only a score or so being capable whips." He added: "That so many women continue to drive and to add their fresh loveliness to the beauty of the Park must be due in part to providence and in part to the police."[24] In 1912 Belle Beach wrote a manual on riding and driving for women. While Beach favored women driving, she opposed them driving rigs pulled by more than one horse. She also favored riding, but she opposed riding astride: "It is only those women who are built like men and very young girls who look at all well astride." Beach clearly wrote for an elite audience concerned with display, noting that gray beaver hats were mandatory for certain forms of driving.[25] Some younger woman favored riding astride, since it was increasingly the fashion in Paris and London and reflected their freedom to "do what they want to."[26]

For wealthy men, riding had earlier become an acceptable pastime. "The first thing, as a general rule, that a young Gothamite does is to get a horse; the second, to get a wife," wrote Charles Astor Bristed, author of *The Upper Ten Thousand,* an 1852 guide to New York elite society.[27] By the late nineteenth century, riding was almost exclusively recreational, practiced especially on bridle paths in large parks of the Olmstedian variety in New York, Brooklyn, Boston, Chicago, Philadelphia, and Pittsburgh.[28] New York alone had nine riding academies with indoor rings for the winter and rented horses available for park riding in the appropriate season.

While Veblen as well as novelists emphasized the status implications of riding and driving, we have seen little indication that the horseless classes resented the spectacles presented in Central Park, in the annual Horse Show, and in the semiannual Coaching Club parades. In fact, their political representatives often supported the provision of leisure facilities for the wealthy and their horses. Numerous lithographs of Central Park show crowds watching the passing scene, and surviving films of elite horse promenades in New York and New Orleans show sidewalks lined with people five or six deep.[29] Their demeanor suggests curiosity more than jealousy. *Rider and Driver* reported large turnouts of spectators to watch Washington's elite promenade from the Ellipse to the river in 1894, and in the same year it pointed out that the sidewalks in New York were jammed for the start of an amateur coaching run.[30] The public wanted to see the latest in horses, the latest in carriages, and the latest in women's fashions.[31] Increasingly, this was also true of the horse shows. The writer of an article on the 1898 Horse Show in Madison Square Garden noted, for instance, that the crowd "was less distinctively aristocratic than it had been in recent years" and that it consisted "of the people who love horses for the horses' sake."[32]

Sleighing

In cities where the climate permitted, sleighing was a popular winter pastime. In Brooklyn, the sleighing season ran, on average, seven to ten days a year. The longest sleighing season in New York City was probably the winter of 1836–37, when snow covered the streets for ninety days. The frequent freeze/thaw cycle led the *Brooklyn Daily Eagle* to issue reports on which streets were suitable for sleighing and which were not. On good days the paper would wax poetic over a snow-covered road, "a broad white, gemmed carpet."[33] New York City sleigh owners often used political clout to block street clearing for a long time. New York City did not attempt to clear snow from Broadway until 1859, while Fifth Avenue was cleared for the first time in 1893.[34]

The *Daily Eagle* covered sleighing as an upper-class pastime, reporting, like a society page, on which prominent Brooklynites were seen on the road, what type of rigs they had, and what kind of horses (often "fancy high-steppers") pulled the sleighs. The paper reported on informal "brushes" (races), on who accompanied whom, and on what attire women passengers or the occasional woman driver wore. Sober business leaders, said the paper, would await snow with much the same anticipation as today's schoolchildren. Snow and the sleighing that went with it had remarkably festive overtones.[35]

Sleighing, however, was not just for the upper class. Grocers and butchers would affix runners to their wagon boxes and employ their draft horses to take their families for rides. One Brooklynite was well known for attaching runners to the roof of a chicken coop, turning it upside down, and using it as a family sleigh in the winter. Livery stables would rent sleighs for a day or two to those who did not have their own horses and reached out to still lower-income groups by offering rides on common carrier sleighs, usually hay rides (stables had plenty of hay). In one case a livery stable sold space on a fifty-seat sleigh pulled by six horses. That particular sleigh broke, illustrating another problem—sleighs were accident-prone. Runaways were common and vehicles frequently rolled over in turns. Racing was hazardous, and collisions could easily occur on crowded streets, sometimes leaving sleighs "smashed to kindling."[36]

For many Brooklyn sleigh drivers, a common destination was the roadhouses at or near Coney Island, where one could engage in drinking and dancing. The taverns competed for this trade, highly valued during their slack season, and decorated their premises with bunting, flags, and even electric lights. They offered prizes to the first sleigh to reach them.[37] Ocean Parkway was a venue popular enough to draw sleighs from as far away as Westchester and Jersey City.[38]

Brooklyn was just one example of widespread sleighing in cities with snowy winters. In Pittsburgh, sleighs were so common that a city ordinance required that "one or more bell or bells be fixed to the head of the horse or horses" drawing the sleigh to give notice of their approach and also forbade them from blocking intersections. One Pittsburgh resident remembered that winter sleighing was the favorite amusement in the nineteenth century and that every "young man of a certain condition" possessed a good horse and sleigh that he drove through the streets, taking "refreshments 'at many an open house.'" A resident of Wilkinsburg, a neighboring town, remembered two-horse sleds "with large, deep throated bells" used for sledding parties. Most "exclusive," he wrote, were one-horse sleighs called "cutters," with fancy bodies painted in several colors and their front bent in the shape of a goose neck, popular in the 1870s and 1880s.[39]

Racing

Horses provided leisure for both spectators and owners through amateur and professional racing. This sport, in both its riding and driving forms, became thoroughly professionalized in the nineteenth century, the first sport to become so organized, although amateur trotting never really disappeared. Except at a few large racetracks in big cities, American racing typically took the form of trotting rather than riding. There were regional variations. Southerners, operating in a different tradition, rarely opposed horseback racing or wagering. In the North, the Puritans had opposed horseback riding as an unseemly display of wealth and also opposed wagering, although races took place in the Boston area as early as the seventeenth century. Republican virtue militated against both racing and betting early in the nineteenth century, and several states, beginning with New York in 1822, banned it. The New York ban, however, seems to have been an exercise in futility, and racing flourished in an economy based on horses.[40]

Promoters hyped the 1823 race in New York City between Sir Henry and Eclipse as a match between the best horses in the North and South. It was a one-time-only match race with huge side bets between the owners of each. Fifty thousand spectators, believed to have bet more than twenty thousand dollars, despite New York State's prohibition, found their way to the Union Course (owned by the prominent politician Cadwallader Colden) in suburban Queens. A Jockey Club dominated by horse-owning Manhattan socialites set the modern racing rules in 1831, and the first racing newspaper, *The Turf Register*, began publication two years earlier. Trotting generally dominated urban racing in the mid-nineteenth century. It seemed less offensive because it was related to everyday uses of the horse and did not have the aristocratic association of riding. Horse racing, however, had come to require professional jockeys and thoroughbred horses descended from expensive, imported British stock.[41]

Owners of trotting tracks argued that racing improved the breed and sought to anthropomorphize their animals, publicizing, for example, the rags-to-riches story of Lady Messenger, who had risen from pulling a butcher's cart to becoming a record-beating trotter. Frank Herbert, the British writer largely responsible for the importation of many outdoor pastimes to the United States, noted that trotting was "the people's sport, the people's pastime, to be supported by the people."[42] One scholar estimates that there were seventy professional urban trotting tracks at their peak in the 1870s, when the creation of the National Trotting Association marked the triumph of commercialization. Ownership of tracks rested with an odd combination of politicos, often Irish immigrants, and members of the wealthy "horsy" set. Thoroughbred racing enjoyed a revival in the 1870s and would ultimately become the dominant mode. The racetrack reflected a democratic culture—anybody could get in and anybody could place a bet. It seems to have had a strong working-class following, especially among the Irish. It is impossible to sort out the extent to which the appeal grew out of the aesthetic admiration of horses and the extent to which the appeal was gambling. Although baseball replaced horse racing as the most popular urban pastime in the 1870s, racing remains a major sport even today, probably because of the attendant gambling.

Amateur racing remained popular as well, although it is difficult to track the information about amateur races. A work by two enthusiasts tells about the amateur racing clubs in Boston and its vicinity. The city had waived speeding limits on specific suburban roads for a long time, most notably on the roadway (completed in 1857) atop the milldam that separated what is now the Charles River Basin from the now completely filled Back Bay. This road, now the Back Bay portion of Beacon Street, was set aside for informal racing ("dust-ups" they were sometimes called). As residences filled up the area alongside the roadway, the municipality applied a speed limit of seven miles per hour. Drivers then shifted to the newly completed Commonwealth Avenue, another suburban artery. In 1890 residents along that now settled street sought a speed limit from the city council.

Surprisingly, the representative from wealthy Back Bay favored the limit. Its leading opponent was Alderman McLaughlin from largely blue-collar, Irish South Boston, who commented: "It strikes me that when a man in the city of Boston has got a good horse, the city ought to provide some place for him to drive." Commonwealth Avenue, McLaughlin added, was virtually the only place within ten miles of Boston where "a man with a fast horse can go have a brush of a quarter of a mile."[43] Since McLaughlin's constituents were unlikely to own fast trotters, he was probably thinking of their interests as spectators. Later drivers would get the Metropolitan District Commission to build a limited-access road along the suburban reaches of the Charles River, the Charles River Speedway. Commercial vehicles were banned from the speedway, and central lanes were isolated from the rest of the road by a railing. The central lanes allowed informal races and high-speed travel without limits. The road served as both a pleasure drive along the Charles and a raceway for the more adventurous.[44]

Boston had constructed a linear park (the Fenway, primarily a promenading park), designed by Frederick Law Olmsted, in the 1880s. Olmsted designed the park's interior drives to exclude views of the city and planted it to resemble an English country landscape. Olmsted deliberately curved its roadways to keep driving at a leisurely pace, akin to Central Park. This did not appease more sporting carriage owners, who wanted a racing venue. The Dorchester Gentlemen's Racing Club convinced the city to build a racetrack in Franklin Field near the Fenway's terminus in 1909, largely because traffic had grown too heavy on Blue Hill Avenue, another artery that the city had previously allowed them as a raceway. The club's annual meets drew up to nine thousand spectators, with mayors John "Honey" Fitzgerald and James Michael Curley driving on opening days. (Both won their heats!) The number of spectators and the participation of mayors who appealed politically to the city's emerging blue-collar Irish electorate suggest that these events provided spectacles for the public as well as adventure for enthusiasts.[45]

Brooklyn followed a similar pattern. There were at least four driving clubs in the city. The largest of them, the Parkway Driving Club, whose members used the Ocean Parkway for informal races, included well-placed politicians who lobbied for maintenance of the road and opposed an el that crossed it. The club built its own half-mile racetrack when the city began to crack down on racing, deliberately seeking a site along the parkway close to the built-up portion of the city so that its members' horses would not tire out on the way to the track. It also shared the racing season with four other amateur, suburban clubs. Three professional racecourses also functioned in nineteenth-century Brooklyn, and one Ocean Parkway roadhouse operated a half-mile track for patrons.[46]

The president of the Parkway Driving Club, Henry Boody, the son of Brooklyn's mayor, observed in 1892 that, "Wherever men have been brave and strong, they have had as their companions, the noblest of animals, the horse." Developing the speed and endurance of horses would help form individual character, "tone the spirit and give stronger and more robust character to manhood."[47] Note the emphasis on the horse as a source of beauty and a model of courage. The real interest of club members, however, seems to have been racing and gambling. Otherwise it would be hard to explain their interest in the latest racing technology, like lightweight sulkies with ball bearings and pneumatic tires. Their track rapidly became professionalized, as gambling among spectators and the scandals that often accompanied professional bookmakers appeared.[48]

In 1889, Pittsburgh acquired the three-hundred-acre Mount Airy property from the heiress Mary Schenley as a gift, created an urban park bearing her name, and purchased more land to enlarge the park to over four hundred acres. The park was developed on Olmsted-like lines, with sculptured landscapes and facilities to support elite leisure activities, such as driving and bridle paths and a golf course. Schenley Park also possessed an Oval, where horse races were often run. These races were held under the auspices of various elite clubs, such as the Schenley Matinee Club and the Crafton Club, and could attract the attendance of several thousand people.[49]

The pattern in Pittsburgh, Boston, and Brooklyn was followed elsewhere, as was shown in an 1884 travel book on American cities. In Buffalo, drives in Olmsted parks were set aside for promenading and Delaware Avenue was reserved for informal races. Cleveland removed speed limits on Euclid Avenue, and Chicago on Drexel Boulevard, so wealthy trotting aficionados could entertain the masses with their races. The travel book noted similar patterns for Savannah, New Orleans, and San Francisco's Golden Gate Park. In Washington, D.C., the Ellipse, an oval road behind the White House, matched the dimensions of a half-mile racetrack. President Grant exercised (and likely raced informally) his prize trotters there.[50]

In the nineteenth century New York City set aside the upper reaches of Third Avenue for country drives and informal races and, as has been noted, Fifth Avenue and the drives in Central Park were centers of promenading.[51] When Manhattan became more crowded, the scene shifted northward; Riverside Drive (1879) and the Harlem River Speedway (1897) became the new centers for the pastime, with Riverside Drive serving more for display and the Harlem River Speedway for informal racing. The emerging residential borough of the Bronx was planned around a "Grand Concourse," with isolated central lanes for higher speeds and outside lanes for promenading and delivery wagons.[52] The Bronx also had a series of Olmsted-like parkways. Whether park road, boulevard, parkway, or speedway, cities laid light gravel pavements, which were unsuitable for heavy commercial vehicles, even if regulations did not ban them. There seems to have been little dissent to public policy measures that provided driving parks and speedways or waived speed limits on some suburban drives, so non-carriage-owning classes may have seen all this as a form of public entertainment worthy of some public subsidy.

Access to Suburban Resorts

Although horse shows and promenades may have provided the middle and working classes with a leisure activity based on viewing, the horse also had a major role in giving these groups access to recreation. By the 1830s, Americans were increasingly willing to see leisure spent in fun and amusement as a positive rather than a negative element in life. By the 1850s, the "cultural transformation of leisure was already well advanced."[53] An important part of the expansion of leisure activities was the growth of parks and resorts of various kinds.

Typically, this change occurred first with the development of rural cemeteries that also served leisure functions.[54] They were built at the outskirts of cities, usually in locations that were accessible only by horse. When Mount Auburn Cemetery outside of Boston was founded in 1831, General Henry A. S. Dearborn, president of the Massachusetts Horticultural Society and one of the cemetery's founders, predicted that it would be "a holy and pleasant resort for the living . . . one of the most instructive, magnificent and pleasant promenades in our country . . . it will attract universal interest, and become a place of healthful, refreshing and agreeable resort." Initially Mount Auburn became more of a resort than its founders had anticipated, with people on horseback disrupting its quiet nature and damaging the park facilities, leading the corporation to impose limits on visitors. Nonproprietors, for instance, had to leave their horses and carriages at hitching posts outside the gate, and omnibuses and excursion wagons were banned from the grounds. Still, traffic was heavy, generating many complaints of fast driving.[55]

Like the cemeteries, urban parks (such as Central Park, Prospect Park in Brooklyn, Schenley Park in Pittsburgh, and Fairmount Park in Philadelphia) attracted more visitors from a variety of classes as they grew older. The middle and working classes were drawn by activities including winter skating, free Saturday afternoon band concerts, boat rides, and strolls along the parks' many paths and gardens. Tropical plants in greenhouses and rides on Swan Boats, as in Prospect Park and Boston's Public Garden, offered more exotic experiences.[56]

How did those without carriages and horses access the cemeteries and parks? Some walked long distances, and by 1870 Central Park's gatekeepers counted more park visitors arriving by foot than by horse.[57] Many others, however, took public conveyances such as omnibuses or horsecars or perhaps rented a horse and carriage from a livery stable. Omnibuses began running from Cambridge to Mount Auburn in 1838. In 1856, twenty-five years after the initial interments, the Cambridge Horse Railroad, the first in New England, built a line from Boston through Cambridge to the Mount Auburn Cemetery gate, and in 1863 a second line brought visitors via another route. The Cambridge Horse Railroad claimed that it made 175 trips a day. The Mount Auburn corporation actually provided free tickets on the line to its trustees. In Pittsburgh the Citizen's Passenger Railway served the Allegheny Cemetery, the city's first rural cemetery.[58]

On weekends, especially Sundays, the horsecar lines often carried their heaviest passenger loads. Central Park attracted one-quarter of its visitors on Sundays. Working-class visitors constituted only about an eighth of the total of annual visitors to Central Park but furnished a much higher proportion on Sundays, the most popular day for park visits. Much depended on the special attractions being offered in the park. In Philadelphia and probably in other cities, the local horsecar company funded Sunday concerts in the parks. New York omnibuses and streetcar lines would fly flags on their vehicles to inform potential customers when the Fifty-ninth Street pond was frozen over. Increasingly, many more visitors were members of the working class.[59]

Horsecar lines also stimulated the growth of various commercial resorts and picnic areas on the urban fringe. In Brooklyn, several "pleasure" parks benefited by connections with various horsecar lines. The Myrtle Avenue Park had a "capacious hotel," dancing platforms, refreshment and ice cream saloons, band concerts, and a "shooting house" for rifle tournaments, and it was utilized by social groups for picnic purposes. In 1861, the Brooklyn Schuetzen Corps, the Caledonians, the Comus Union, and various parochial schools visited Myrtle Avenue Park. Morris Grove, a "delightful shade retreat for health and amusement," was about two miles west of Jamaica, Queens, and on a rail line on which cars traveled hourly from South Ferry and Jamaica, stopping on signal. The Bath Hotel, in Bath, Long Island, advertised in 1863 that it could be reached by taking the "city cars to Greenwood" and then stages to Bath.[60]

In many ways the most publicized play area in the nation was Coney Island, which had commenced its history as a beach resort several decades before the Civil War. Originally not easily reached by land from Brooklyn because of intervening salt marshes, it was served primarily by steamboats from Manhattan. In 1850, omnibuses began to operate on the Coney Island Plank Road, and the development of horse-operated rail transit in Brooklyn in the mid-1850s created even better access. As one letter writer to the *Brooklyn Daily Eagle* noted in 1859, "No one can doubt that if a first-rate horse railroad were built, connecting the city and the sea-shore by an hour's ride, thousands would go daily to enjoy it, the ocean scenery and the purifying and health-giving sea bath."[61] In the early 1860s Coney Island and other recreation spots attracted at least four more lines. One charged twenty-six cents for a trip covering the 10.2-mile-long road from downtown Brooklyn to Coney Island and operated twenty-eight open cars; as many as 236 horses were used in the summer. With the help of the horsecar, Coney Island had become what the *Eagle* called "the most popular resort of a great city . . .

[and] emphatically the people's bathing place . . . within the reach of the humble street."[62]

The importance of horsecars for Coney Island was reflected in another *Brooklyn Daily Eagle* piece dealing with the first day of the 1870 season—Sunday, June 13. The reporter wrote that the Coney Island Railroad cars had been crowded all day and took care to comment on how well prepared the railway had been for the anticipated crowds. The road's superintendent, he said, had ensured that "the season was opened on that road with commendable éclat," and the cars ran "as regular as clockwork every twenty minutes," while the "ride along the road is really delightful." In addition, the article mentioned the prominence of horses in the resort's attractions, including the showing of "blood horses" and the holding of informal horse races. A mile stretch of road along the route to the resort was said to be a "favorite ground for 'a brush,' and every afternoon many hundreds of dollars are spent, and much wine is drunk on the result of impromptu races got up on this road."[63]

Profaning the Sabbath

Equine-powered transport of leisure seekers to resorts on Sunday, however, incurred opposition, since orthodox Sabbatarians sought to protect that day from the forces of secularization and modernization. While the nation was undergoing a loosening of the bonds of religious orthodoxy, much resistance from traditional denominations persisted, and Blue Laws remained on the books in many cities and states. Liberal Protestants argued, however, that Sunday was a day for men (and women) to enjoy culture, with its uplifting possibilities.[64] Street railways, whose horses allowed the masses to seek modern, secular entertainment, were an obvious target for Sabbatarians. Disputes over running them on the Sabbath often had nasty nativist overtones, since many members of America's emerging ethnic communities, especially German Americans, enjoyed holding picnics and taking Sunday excursions that often included much beer drinking. Many ministers objected to what they considered the profanation of the Lord's Day, demanding legislation banning horsecar operations, especially to resorts and parks on Sundays. Major conflicts occurred in Baltimore, Brooklyn, Philadelphia, and Providence when they built their first horsecar lines in the 1850s.[65]

Overturning the ban on Sunday horsecar operation in Brooklyn was especially contentious, with the battle lasting three years in the mid-1850s. When the Brooklyn Common Council had granted the City Railroad permission to run its horsecars on the streets in 1853, it had also forbidden the company from operating on Sunday. This restriction was rationalized as being in tune with Brooklyn's reputation as the "City of Churches" and "a moral and church-going community." In January 1856, a group of Brooklyn citizens petitioned the Common Council to compel the City Railroad Company to run cars on Sunday to the Eighth Ward, a suburban district. The petition argued that many of the petitioners had recently moved to the Eighth Ward but still retained pews in their old neighborhoods—thus, a Sunday car would enable them to attend church on the Sabbath. In addition, the petition maintained that the lack of public "conveyance" on Sunday discouraged newcomers from "locating" in this new district, limiting its growth and development.

The petition galvanized to action the advocates of strict Sabbath observance. Letters of complaint bombarded the *Brooklyn Daily Eagle*, and petitions defending the restriction flooded the Common Council. These petitions maintained that Sunday cars would violate the Sabbath, be "detrimental to public morals," and make it easy for "a large, dangerous and ideal floating population [from] New York to destroy the present quiet, peace and reputation of our city." While some Eighth Ward residents favored Sunday cars, others opposed them because they feared that they would make their neighborhood "intolerable as a place of residence, and as a consequence property would materially depreciate in value." Sunday operation, they claimed, "would open wider the . . . channel for the lower strata of New York to pour in upon our glorious suburbs . . . to indulge in garden robbing and gunning exploits, enlivened by a dog fight or a drunken bout." The fact that ministers of elite churches and their parishioners appeared in the forefront of those opposing Sunday streetcars highlighted the class dimension of the dispute. The wealthy, opponents of the ban charged, could attend church services in their carriages, whereas the working class and even the middle class depended on public transportation.[66]

The Railroad Committee of the Common Council took up the issue of Sunday cars in mid-February 1856 and heard petitioners on both sides of the question. Those opposed were most vociferous, as religious arguments were combined with the fear of injury to property values if the "worst classes" of New York and Brooklyn ("riff-raff") were allowed to flood the Eighth Ward. Petitioners cited the effect of Sunday railroads on Harlem, where "crowds . . . plundered gardens, and fruits, and destroyed those things which made a home in the suburbs habitable." Without saying whether it agreed with these arguments, the committee maintained that it could not recommend eliminating the ban, since the streetcar company had signed a bond that forbade them from running Sunday cars. The issue remained largely out of public discourse until the election of a new mayor, Samuel S. Powell, who took office in January 1857.

Powell, who was interested in Brooklyn's residential development in competition with New York City, argued in his annual message that the ban on Sunday cars was hampering the city's growth, especially in the outer suburbs. By making it difficult for "many persons in moderate circumstances from becoming residents of the outer Wards," the ban had "a depressing influence on the value of property." The city railroads, he argued, provided "utility and convenience" to city residents. Brushing off fears about the threatened "evils" of Sunday cars, he noted that riders of New York City's Sunday horsecars were "as quiet and orderly as any class of our citizens." He finally persuaded the Common Council to approve horsecar operation in 1857.[67]

The wild "scenes of riot and dissipation on the Sabbath" predicted by some ministers if the ban was removed, noted the *Brooklyn Daily Eagle,* had proven to be false. In fact, noted the paper, "Sunday recreation seekers are a most orderly and respectable class, composed chiefly of the hard-working population—people of industrious habits who stick to their occupations during the week and avail themselves of their day off to escape from the scene of their toil and get a glimpse at the world beyond."[68]

Horses had clearly facilitated the ability of middle- and working-class people to access places of recreation where they could spend their day of leisure in enjoyable ways. Eventually churchgoers rode horsecars to services, and soon some ministers were attempting to persuade the streetcar companies to put more cars on the lines that ran to their churches. Six to eight cars were supposedly filled every Sunday with attendees at Henry Ward Beecher's Brooklyn church. Beecher observed in an 1870 sermon that he could have gotten many signatures in his church in the 1850s against Sunday cars but none "to stop a rich man from riding in his coach on Sunday." Development and access had triumphed over traditional morality—horses had to work an extra day each week to facilitate human churchgoing and leisure.[69]

The urban horse thus had a liberating influence, opening a new world of leisure for nineteenth-century urbanites of all classes, most obviously by allowing access to either the real countryside or the large parks and cemeteries that pretended to imitate nature. New commercial amusements appeared: the roadhouse, the seaside beach, and the racetrack, all of which would have largely been out of reach of most urbanites without horses adapted to urban applications with their mechanical attachments. Not many working New Yorkers could have reached Central Park, beyond the city fringe when opened, without horse transport, nor resorts such as Coney Island. The same applies in other cities. Speed, whether through direct participation in driving or vicariously at the racetrack and parkway, also became part of urban leisure patterns. Finally, the horse was an aesthetic object, beautiful in motion in a variety of gaits, as was the teamwork between horses. While a rider or driver might enjoy this best, certainly spectators appreciated it, too.

CHAPTER FIVE

Stables and the Built Environment

Historians usually cite the Iroquois Theater fire in Chicago in 1903, which took 603 lives, as the deadliest of urban fires, but on May 27, 1887, another conflagration had taken the lives of 1,185 New Yorkers, all male.[1] The victims of the New York fire, like many inarticulate urban workers in the nineteenth century, left no historical record. These recent urban arrivals (there were probably very few native New Yorkers involved) died an extremely painful death, but the public displayed little pity; rather, people wondered how long it would take to replace the victims. Nobody conducted a memorial service or offered them a decent burial. In fact, health authorities ordered their remains hastily dumped into the Hudson River.

The New York fire has gone unnoticed by historians because it victimized a largely forgotten but major part of the nineteenth-century workforce—draft horses. At 1:30 in the morning, a fire had broken out in the three-story, brick stable of the Belt Line Street Railway (part of the Central Park, East and North River Railroad Company) on Tenth Avenue between Fifty-third and Fifty-fourth streets. The company stabled 1,230 horses there, of which night watchmen saved only 45. Despite the brick walls, the building was highly flammable, with wooden frame, stalls, and floors. Moreover, the stable contained more than four thousand bales of hay, five thousand bales of straw, and twelve thousand bushels of grain. The conflagration evidently started in a paint locker in the first-floor room where the Belt Line Railway repaired its rolling stock (154 cars also went up in smoke).

There does not seem to have been any attempt by the few rescued horses to rush back to their stalls or any "freezing" by those being rescued, a common problem in stable fires, probably because the flames spread too rapidly (the whole stable was gone within a half hour). The newspaper account described the "pathetic whinnies and cries" of dying horses locked in their stalls. The narrative also noted the heroism of the security guards and praised the courage of the New York fire fighters. They prevented the fire from spreading beyond a few nearby buildings, which housed about thirty families, so that only one human life was lost. Bizarrely, the stable's five cats somehow survived and came crawling out of the wreckage the next day.[2]

The fire graphically indicated the problems of relying on the horse as a prime mover and stimulated a "growing public sentiment that big stables in a crowded city are dangerous nuisances."[3] The fire also illustrated one of the reasons why city residents did not want to live near stables. In theory, horse power was made up of thousands of independent sources not subject to systemic failure, like a twentieth-century power grid, but this fire and numerous epidemics proved otherwise. Systemic failure could and did occur. Stables were perhaps the weakest link in the system, since all-too-common stable fires and epidemics of contagious diseases, which spread as rapidly in densely populated horse stables as in densely populated human tenements, could disrupt vital power to entire neighborhoods, even entire cities.

The Belt Line Street Railway served Fifty-ninth Street and both Manhattan waterfronts on a circular route. Those neighborhoods lost transit service for an extended time.[4] The *Times,* as angry as most New Yorkers over the failure of equine power, editorially called for a form of "mechanical traction" to replace horses—the living machine. Reflecting this reality, the executive committee of the Belt Line met on the day after the fire to consider their options. Reportedly, the promoters of cable, electric station, and independent electric motor systems had been in touch with company officials since the fire. The executive committee concluded that "this was a good time to introduce another motor." Thus, as the *Times* observed, "the fire is likely to be productive of an innovation in street car motive power."[5] At a time of rapid change in the technologies of urban transit, this was not an unrealistic position.

The Ecology of Stables

The horse in its role as a living machine, as we have discussed, shaped the ecological patterns of American cities, but cities also possessed their own ecology relating to horses' metabolic and housing needs. The urban built environment reflected the city's dependence on horses primarily in the character of street paving, the distribution of retail and wholesale markets selling provisions for horses, and the location and construction of stables. In this chapter we consider the latter element—the construction and distribution of various types of stables, as well as the problems stables created for their neighborhoods and for the city as a whole. Horse ecology both mirrored and differed from human ecological patterns, responding to commercial and personal needs for transportation but also being shaped by regulation. Stable patterns in many cities showed common features, with each city's economic and spatial characteristics determining variations. Major variations seemed to exist between very dense eastern cities, with tight topographical boundaries, and western cities with lower densities.

The housing of horses was a major issue for all crowded cities, although horse occupancy per stable varied greatly from locale to locale. In 1900, Boston had the most horses per stable in the nation, at 7.8, with New York close behind at 6.7,[6] and Pittsburgh and San Francisco tied for third at 4.8. The cities with the smallest number of horses per stable were largely in the Middle West or Far West, with Cleveland, Detroit, Denver, Indianapolis, and Los Angeles all having fewer than 3 horses per stable.[7] Average stable size almost doubled in many cities between 1900 and 1910, continuing a trend toward the construction of larger stables that had begun in the last decades of the nineteenth century.[8]

The Report on the Sanitary Condition of the City, issued in 1866 by the New York Council of Hygiene and Public Health of the Citizens' Association, provides detailed information about the city's stable inventory just after the Civil War. Stables were concentrated in numerous streets and neighborhoods, came in all shapes and sizes, and were constructed of both wood and brick. Some, like the "well-known Bull's Head stables," located on Twenty-third Street between Lexington and Second avenues, were quite large, containing one thousand stalls. Other districts had a mix of small and medium-sized stables. The city's Third Sanitary District was located in the lower part of Manhattan Island and had 108 stables, housing 585 horses. Of the 108 stables, 68 contained fewer than 5 horses (202 in total), while 40 contained 5 or more horses (383 in total). As the city grew, so did the number of stables, especially those that provided cab and livery services. The livery firm of Ryerson & Brown, established in the 1830s, was worth $250,000 by 1882 and had stables scattered throughout the city.[9]

Somewhat different patterns existed in Boston, the city with the highest occupancy of horses per stable. To get an accurate number, we counted all the stables in Boston from the 1867 *Sanborn Fire Insurance Map.* Fire insurance maps required completeness, since underwriters consulted them before setting fire insurance rates. The maps were probably accurate, as suggested by the wealth of detail that they contained and the powerful economic incentive for accuracy. Stables are easy to spot—the mapmakers put a big X through each one, since they posed a well-known fire hazard, not just to themselves but also to their neighbors. They were built from wood and were full of highly combustible straw and hay, and neither species of occupant was very safety conscious. On the first floor, sta-

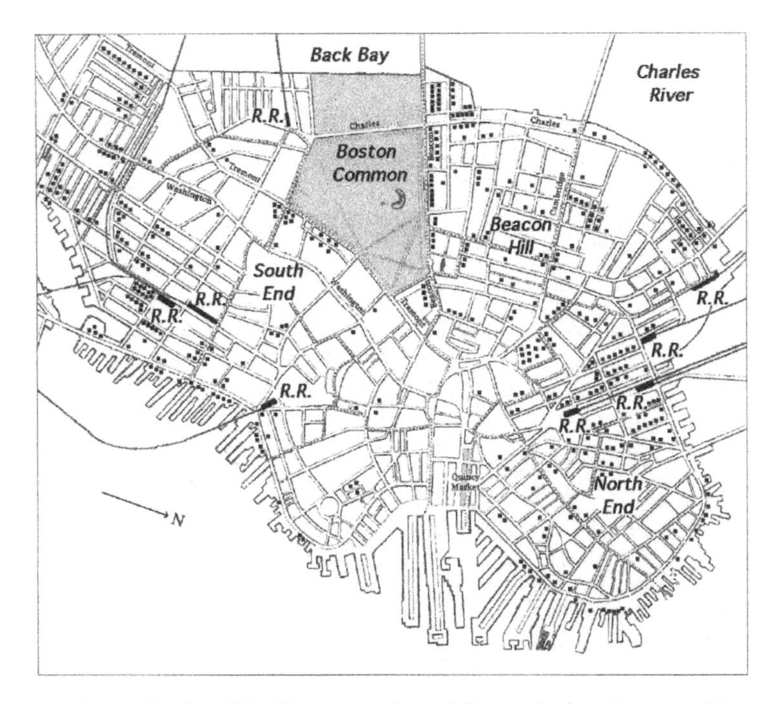

Figure 5.1. Boston Stables, 1867. This map is derived from a *Sanborn Insurance Map.*

ble managers usually stored vehicles and their repair shops, and these contained such highly fire-accelerating materials as paint, varnish, and flammable oils. Contemporary accounts suggested that animals kicking over kerosene lamps caused many fires, like Mrs. O'Leary's famous cow.[10]

The 1867 *Sanborn Map* showed 367 stables in Boston. Just over a third were in back lots behind other buildings, and most of the other stables were on blocks that included more than one stable. Nearly two-thirds of all stables were two stories tall, while one-quarter, mostly small one- or two-stall back lot sheds, had only one story. All of the others were three stories tall, with the exception of one building with four floors. Even those rare stables not built out of wood had wooden floors and frames.

Figure 5.1 illustrates Boston stable locations.[11] In Boston, a maritime city, more than a third of all stable owners had sited their facilities within a block of the waterfront. In many cases the stable was actually on a wharf, where the horses were probably used for hoisting as well as hauling. Nearly as many stables were located near Boston's four railroad terminals, and almost all of the others were on or close to the four main streets (Tremont, Washington, Beacon, and Cambridge), which provided Boston's only land links to its hinterland.

Location also varied by function. Express companies and the private stables of warehouses or factories were found near transportation hubs. Street railways tended to build their stables close to the end of their routes, where land costs were cheaper. Moreover, they dispersed their horses among a large number of barns. At its peak the West End Street Railway housed its eight thousand animals in sixty different stables, many beyond the city limits. This pattern cut the number of dead hauls needed and reduced the risk of losing the entire herd to fire or disease.[12] Livery and boarding stables serving the well-off were clustered near Beacon Street and Tremont Street, both of which led to picturesque suburban drives. Finally, there were some private stables on wealthy Beacon Hill (probably small stables that predated city regulation or whose owners could ignore regulation) or in the working-class North End (likely the illegal installations of peddlers, independent carters, or cabbies).

The *Boston City Directory* provided a woefully inadequate count, showing only 105 stables in 1870 and 209 in 1900, likely an undercount by a factor of ten. Most horses lived in private stables owned by users like the city's 134 milk dealers, 826 express companies, or 53 breweries. Livery and boarding stables clustered together in locations like River Street on Beacon Hill for other reasons than access to roads to the suburbs. There were probably economies of agglomeration, such as easy comparison for renters seeking the "perfect" stable or access to support services like feed companies, blacksmiths, and veterinarians. Also, public opinion led to regulation through the building permit process, which produced clustering.

The following list gives information on the location of Boston stables in 1885, reflecting changes since 1867:

Waterfront (within one block)	122
Beacon Hill	97
North End	56
North Station (R.R.)	33
South Station (R.R.)	58
Beacon Street	41
Tremont Street	29
Washington Street	40
Cambridge Street	39

Fewer than 4 percent of Boston's buildings were stables. Of those, 37.3 percent were at the back of a lot. Seventeen percent were three stories tall, 59 percent were two stories, and 24 percent were one story. There was one four-story stable.

The 1885 *Sanborn Map* showed roughly the same number of stables (385) as had been present twenty years earlier (367), despite the near tripling of the number of horses in the city. Stables were larger, and the rules governing where stables could be located were probably better enforced. (There were very few stables in the newly developed, elite Back Bay, except on upper Newbury Street, which had been set aside for that purpose.) Very large stables were built in the newly developed neighborhoods of the South End and South Boston. There were no small wood stables in the district burned out by the 1872 fire, and the number of small back alley stables declined elsewhere. The consolidation of stables reflected not only a more stringent building permit process but also the centralization of horse-using businesses, as large firms squeezed independent teamsters out of business. Building regulations that required expensive, modern stables also helped force out the small entrepreneur. Some companies could and did move their stables to the suburbs, where regulation was less stringent.

Pittsburgh showed patterns similar to those of Boston in that it had a relatively low ratio of horses to humans but a high ratio of horses per stable. Bodies of water, limiting their land areas, confined both cities. Pittsburgh's larger stables clustered in the downtown area and along the river flood plains, the location of both flatland and railroads. Smaller stables that served individual owners were scattered throughout the city. During the 1890s, stable construction constituted 5 to 7 percent of new buildings in the city. Figure 5.2 shows the location of major livery stables in Pittsburgh in 1892.

Stable Construction and Design

Residences for horses, like residences for humans, ranged in quality from poorly constructed wooden sheds to large brick and iron structures utilized by horsecar and express companies. As a major part of the urban built environment, stables were constructed from available building materials and ornamented as fashion suggested and owners could afford. Because they came in all shapes and sizes, we will discuss them in different cities from the perspectives of size, construction, and function.

As has been noted, those operations that had the biggest stables were the firms providing transportation and delivery services within the city. Because horses represented a capital investment, firms had an incentive to maintain their horses under conditions that protected their health and longevity. Streetcar companies required the largest stables not only because they owned more horses but also because they had to accommodate hay and grain storage and provide shoeing,

Figure 5.2. Pittsburgh Livery Stables, 1892.

harness repair facilities, manure bins, vehicle storage, and even horse hospital facilities. The companies worried about diseases growing out of overcrowding and tried to reduce these conditions. Writing in 1870, for instance, Robert McClure, M.D., V.S., noted that "city car" or "railroad stables" were a great improvement over the "stage and omnibus stables of the past" and that their "superior ventilation, light, and cleanliness" had "almost banished glanders and farcy from our midst." He noted, however, that the best conditions existed in the stables of the large rather than smaller and "poorer companies."[13]

Transportation firms often built brick and iron-framed stables even if regulations permitted cheaper construction. Wood floors remained, since they were easier on hooves. The typical streetcar stable was three stories tall. Usually, the top floor held grain or straw, while horses took a ramp (or, rarely, an elevator) to the second floor and vehicles were kept on the first. Having horses on the upper floors, however, as the Belt Line Street railroad discovered, made it impossible to rescue them in the event of a fire. Placing horses on the first floor and vehicles on the second created a different problem. Fumes from the horses' urine, especially when it bled from the floors on a hot summer day, were strong enough to blister the paint on vehicles stored on a floor above.[14]

Specific information about stable design and improvements, relating especially to the care of horses and fire prevention, is available for several large stables in New York City constructed in the late nineteenth century. New York's stables may have been exceptionally crowded. At the supposedly normal stall size of thirty-six square feet, New York's horses would have occupied 1.6 million square feet of the most expensive real estate in the nation.

In 1876, the New York & Harlem Railroad constructed a new $150,000 stable on Fourth Avenue between Thirty-second and Thirty-third streets that adjoined its older stable and depot. The *New York Times* called it "A Model Car Stable" and "A Triumph of Ventilation and Drainage." The two-story brick stable was 182 by 172 feet. It had 273 stalls on the first floor, with grain and hay stored on the second floor. The stable contained 906 stalls and a hospital that could accommodate 75 horses. Its drains were flushed with water every day and disinfected once a week. The stable employed fourteen blacksmiths, twelve carpenters, four harness makers, two engineers, and about one hundred hostlers.[15]

The New York City Second Avenue Railroad Company constructed a new depot and brick stable at Ninety-sixth Street in 1878. The three-story building was 475 feet long and 200 feet wide. The exterior was decorated with colored bricks and plaster horse heads. Horses ate and slept on the second floor, while the third floor was used to store feed and implements. A skylight that ran the length of the building provided illumination. At the rear of the building was an "exercise ground for temporarily disabled horses." In addition, the company announced that it would construct on an adjoining lot tenements for its drivers, conductors, and stable laborers, but there is no information as to whether this was ever done.[16]

The ultimate "state-of-the-art" street railway stable, built by New York's Seventh Avenue Street Railway and completed in 1888, a year after the Belt Line fire, held more than two thousand animals. *Street Railway Journal* claimed that it was the largest stable in the world and reported on it no less than four times, a measure of its novelty.[17] It contained twenty-five hundred stalls on four floors and covered the entire block from Sixth to Seventh avenues between Fiftieth and Fifty-first streets. The stable was one of the first New York buildings to contain sprinklers, and its builders added brick stall partitions and fire doors leading to its ramps. The operators substituted peat moss for straw for both sanitation (it absorbed urine) and fire prevention.[18] This may have been the last giant stable constructed for streetcar horses, since electrification of street railways, which would eliminate the largest urban stables, was already under way.

There were other quite large stables. The growth of various package and mail delivery systems created a need for horse accommodations. Both Wells Fargo and the United States Express Company maintained stables in Jersey City serving the New York metropolitan area. The latter's stable held 275 horses. Built in 1885, it was 250 by 100 feet and was made of brick, iron, and concrete, with an iron roof. Horses were stored on the second floor, and the top floor was used for paint shops and harness storage. In spite of its supposedly fire-resistant construction, this building burned in 1888.[19] In 1903, William H. Seaich, owner of the Opera Stables and president of the Livery Stable Keepers' Association, constructed a six-story stable for six hundred horses and the storage of cabs and carriages on Thirty-first Street in New York City.[20]

Companies that relied on horses to make regular deliveries of their products had a special incentive to provide them with good housing and to maintain them in good condition. The horses were a symbol of company pride, important to their sales success, and a major capital investment. In 1910, for instance, the H. J. Heinz Corporation of Pittsburgh issued a handsome, illustrated brochure that proudly described the three-story, fireproof building that housed its delivery horses. Heinz lit the structure by electricity, heated it with steam, and ventilated it with electric fans in the summer. All windows were screened and shaded. As usual, the horses lived on the second floor, with wagons parked on the first floor and feed and bedding stored on the third. The horse stalls were of open pipe, with cork block floors and swinging sides. Each manger had separate, electrically controlled compartments for hay, grain, and water, and horses were fed automatically. Electric cleaning devices were used to curry the horses. A hospital stall was equipped with a Turkish bath stall and a footbath! The brochure boasted that the stables were "one of the most modern, practical and complete commercial stables in the world."[21]

A new urban phenomenon in the late nineteenth century was the department store, located in the hearts of large cities such as New York, Boston, Chicago, and Pittsburgh. As their business grew, department stores developed delivery services.[22] These, too, required stables, ones that were designed with an eye toward impressing the public. In 1875 Marshall Field and Company in Chicago constructed stables near their store that housed about 100 horses and 50 wagons; by 1885 the firm owned 163 horses, and by 1895, 344. The store's management insisted on sanitary conditions and good treatment of its horses. As the business grew, the store provided free delivery to Chicago's growing suburbs, and by 1902 it had established separate suburban stables.[23]

Macy's, in New York City, grew from a fancy dry goods store to a diversified department store between 1858 and 1877. The store acquired a stable for delivery horses on West Fifteenth Street, two blocks away from the emporium. In 1875 the store moved the stable to West Nineteenth Street. Exact information on the number of horses housed in the Macy's stable is unavailable, but it was a large and growing number. By 1872, drivers, wagon boys, and helpers constituted 13 percent of the firm's workforce. Deliveries escalated as the business grew, and by 1896–97 the store was delivering over 2,500,000 packages a year. As a result of this demand, Macy's hired more men, built more wagons, and enlarged the Nineteenth Street stable. Later it constructed a delivery branch and stables at 148th Street at the city's outskirts.[24]

Macy's, like other large New York City department stores, faced serious issues coping with deliveries to rapidly growing suburbs. Initially the store offered free delivery only in Manhattan, Brooklyn, and Jersey City. In 1888, in response to competition, Macy's advertised that it would provide free delivery within one hundred miles of its store for all purchases of five dollars or more and then established delivery routes throughout New Jersey and Long Island. Many, however, were not profitable because long distances increased costs and put great strain on the horses. Enterprising individuals, however, founded suburban express services that delivered packages from multiple stores.[25]

Livery and Boarding Stables

Another major part of the stable inventory of any American city were sales, livery, and boarding stables, which were smaller than the huge horsecar and department store enterprises. At least initially, the livery and boarding functions were blurred. A local stable might rent, board, and even sell a few horses on consignment. In Boston, city directories show that the livery and boarding functions became increasingly separate over time. This was likely to be true elsewhere. Evidently, boarders feared that unscrupulous stable keepers might rent out their horses when the owners were not using them. Throughout the late nineteenth century, the number of livery and boarding stables increased with the growth in city populations. Based on city directory counts, between 1870 and 1900 the number of livery stables in Boston increased from 71 to 175; in Pittsburgh, from 12 in 1866 to 39 in 1879 and 82 in 1892; and in Philadelphia from 78 in 1859 to 121 in 1910.[26] Detailed descriptions of their features, if somewhat exaggerated (usually advertisements), can be found in various city guidebooks and business directories, providing insight into stable conditions and also the varied uses of horses.

Data from Pittsburgh business directories is especially informative. In 1880, for instance, A. Jackson & Sons operated an Exchange Livery Stable (founded in 1850) that occupied a space 150 by 150 feet on 174–184 Penn Avenue in downtown Pittsburgh. The *Pennsylvania Business Directory* for that year noted that the stables were "classed with the largest in the United States [and were] equipped in a most approved and thorough manner." J. L. Kennedy of Allegheny City operated a "comfortable" twenty-horse stable that supplied "plenty of good food" and promised "the best attention . . . by careful and experienced grooms." B. F. Dyer's Livery, Boarding and Sale stable at 612–614 Grant Street in downtown Pittsburgh was "perfectly lighted, and drained and ventilated." And the Grand Central Boarding and Livery Stables, C. A. Warmcastle, proprietor, located in Pittsburgh's residentially desirable East End, occupied a three-story brick building furnished with electricity and an elevator. The stable had first-floor stalls for ninety horses. A special feature of its boarding services was "clipping horses by steam power."[27]

An unusual phenomenon seemingly particular to Pittsburgh involved combinations between livery stables and undertakers. Funerals provided a substantial amount of livery stable business and often provided the margin between profit and loss. Of the fifteen livery stables listed in the Pittsburgh section of the *1880 Industries of Pennsylvania*, eight had "undertaking" listed as either their primary or secondary business activity. The 1892 *Pittsburgh City Directory* listed eighty-two livery stables, of which twenty also provided undertaking services; of the forty-eight undertakers listed, twenty provided livery service.[28] The Livery, Boarding, and Sale Stables of M. F. Leslie & Bro. provides a good example. The firm was located in the Lawrenceville section of Pittsburgh and had forty horses. The firm promised to provide "elegant new broughams, landaus . . . with coachmen in livery, besides top-wagons, surreys, and phaetons for park or road driving." The broughams and landaus may well have been rented for weddings. The firm also made "Undertaking and Embalming a Specialty" and as undertakers supplied "every requisite from the casket and mourning badge to the hearse, coaches, and opening of the grave, in fact take full charge of managing every detail in the performance of the last rites of the dead."[29]

The activities of the Sampson family, operators of the Sampson Funeral Home and related businesses, located in downtown Pittsburgh from 1859 to 1922, provide insights into the history of an undertaking firm that used horses in their business well into the twentieth century. Hudson Sampson, the founder of the firm, expanded his business over the years, moving from the provision of boarding and livery services to the operation of the Allegheny Express Company, the Pullman Taxi Service Company, and, in the early twentieth century, the Auto Livery Company. Sampson was supposedly especially conscious of the quality of his horses— "whenever he saw a horse that would mate little prettier with any of his three or four blacks," said his biographer, "he would buy him and sell the poorest of those he had." He believed that his business was best advertised by "fine vehicles more than anything else—fine hearses, fine wagons, satin-coated horses, and the harness, the driver himself, and every part of the turnouts, spick and span, shining bright." In 1910, his son Harry introduced Pittsburgh's first motorized funeral vehicle, although horses were not completely replaced until some time later.[30]

The use of undertakers' horses for other purposes makes sense, but this combined pattern did not appear in other cities that we have examined. In New York City, livery stable owners formed the Livery Stable Keeper's Association in 1881— a combination to raise rates to undertakers for funerals.[31] The liverymen maintained that competition had driven prices down to the point where they could not make a profit. In addition, they complained that mourners often overloaded their carriages during funerals, straining their horses and damaging the carriage interiors. Liverymen and undertakers finally reached an agreement on rates, avoiding a disruption of New York funerals. Five years later Brooklyn stable owners and undertakers reached a similar agreement.[32] In Pittsburgh, however, the undertakers had broken the control of liverymen over prices by obtaining their own horses and going into the livery stable business themselves.[33]

Information about the management practices of small commercial stables is limited, but the R. G. Dun Credit Reports at Harvard's Baker Business Library provide some information for Boston area stables for the years 1865–95. Dun reported on only medium-sized livery and boarding stables and on less than a quarter of those. Presumably, most stables, especially the smaller ones, relied on informal family/neighborhood credit networks.[34] Most stable keepers leased their facilities, since Dun always noted exceptions. Horses had to be fed, whether used or not, creating a pattern of debt in late winter months, especially to feed dealers. Renting horses was highly seasonal and cyclical, with urban economies expanding in the summer. The nineteenth-century economy was also subject to strong year-to-year fluctuations. Whether renting for business or pleasure, there were more customers in good years and good seasons. To be sure, livery rentals picked up in snowy weather, when sleigh rentals were in high demand, but this also meant that stables had to keep two sets of vehicles on hand. The popularity of sleighing cannot be underrated and the season, even as far south as New York, sometimes lasted ninety days.[35] The spring "courting season," as it was called in the trade, also demanded specialized vehicles offering a modicum of privacy.

The Dun reports suggest some entry and mobility patterns for owners. A blacksmith or cabdriver might rent space in his stable to others. A stablehand with a good reputation might start a boarding stable, which required less capital. Such stables, however, faced the problem that they did not have livestock to mortgage; typically, liens on horses provided the surety for credit from feed dealers and others. Over time, boarding stables might become livery stables.

To judge from the Dun reports, mobility in both directions was common, although the booming nineteenth-century economy probably allowed for more successes than failures. One example is illustrative. In 1876, John Riedle, a stable manager, bought a building for a livery stable at 196 Tremont Street, a good location on one of the main roads out of Boston. He had trouble covering expenses, and two years later Dun reported that he was a "weak and undesirable risk." Riedle recouped, switching for a short time to boarding, a less capital-intensive business. Two years later, he had acquired "a number of good boarders," was grossing twenty-five thousand dollars a year, and had seen the value of his building quadruple.

Most debts of the stable owners covered by the Dun reports were seven thousand to ten thousand dollars. Despite a wave of failures in 1873, most prospered and grew with Boston's economy. Still, difficult years like 1873 or an especially bad winter must have led to the forced sale of some horses, tempted owners to overwork old and sick ones, and produced some underfeeding. Feed dealers held most of the loans, having extended credit in the late winter or early spring. Chronic capital shortages probably explain some of the poor conditions of urban stables, since the industry was well aware of the advantages of cleanliness and "fireproof" structures.

Similar conditions likely prevailed elsewhere. An 1873 *Brooklyn Eagle* article on livery stables reported that there were fifty livery stables in the city ranging in value from five thousand to thirty thousand dollars. Most, however, were at the lower end. Some stables provided boarding facilities, as well as renting teams and also doing a general trucking business. Sleighs for winter rental were available. One large, three-story stable described in the article had accommodations for one hundred horses and numerous vehicles.[36]

Private Stables

The condition of small, private stables—that is, stables owned by individuals to provide shelter for their horse or horses—could vary greatly. Individual businessmen, such as peddlers, scrap collectors, or draymen who owned a single horse, maintained many. There were large numbers of ephemeral one-horse stables, including many warehouses with a small stable inside and many tucked away unnoticed in industrial zones. In addition, since many were adjuncts to other business, they showed up in neither census reports nor city directories. Health boards and housing reports sometimes tried to track them, but, by their own admission, their records were inaccurate. Small, private stables also varied in quality from neighborhood to neighborhood, usually depending on how much stables were perceived as a threat to health. And, of course, there were the usually high-quality stables maintained by the few wealthy urbanites who could afford to own their own horses for recreational purposes.[37]

Descriptions in various sanitary surveys provide us with a picture of stable conditions. The 1866 Citizens' Association of New York survey, for instance, described stables in poor areas that were "crowded [and] frequently neglected and uncleanly," located in either basements or the rear of buildings, kept in "slovenly condition" and serving as "fever-nests."[38] Some very small stables did not even have windows because their owners did not want the expense and thought that windows might make entry easier for a thief.[39] Greenwich Village held 196 stables, mostly small back alley shanties. The desperately poor, very densely populated Lower East Side contained only 288 stables among its 4,400 buildings, with many in the basements of buildings, where horses and humans shared the same entrance. One inspector reported that "the prevalence and fatality of pulmonary diseases among horses in crowded and neglected stables is only equaled by the fatality of like maladies in the women and children of tenant-houses."[40] One problem noted on West Twentieth Street was the close juxtaposition of stables and rear dwelling houses, so that horse odors penetrated into bedroom windows

Chicago had many more small stables than did New York, Boston, or Pittsburgh, probably because of the city's territorial expanse. In the 1880s, the *National Livestock Journal* estimated that three-fourths of the city's stables were held for private purposes.[41] Several of these were in deplorable condition, even shared by humans and animals alike. On February 3, 1892, the *Chicago Daily News* interviewed a peddler, Michael Holeran, who shared his basement home at 227 West Fifteenth Street with his horse, his wife, and his two children. The horse had its own room "neatly carpeted with straw." Holeran claimed that he "was willing to sleep with a clean beast like that himself," even though he stayed in another room. Holeran's upstairs neighbors complained about the "unwholesome odors" and blamed a family illness on them.[42]

Nine years later, in 1901, social worker Robert Hunter surveyed tenement house conditions in Chicago, including the city's stables. Hunter found that the Holeran's living conditions were not so unusual, commenting that "unmarried Greeks frequently share their own rooms with their horses and Italians often stable them on the lower or basement floors of their tenements." Hunter noted that Chicago's laws, unlike New York's, allowed horses and humans to share the same buildings. In a single West Side district of tenement houses and industry, Hunter found 537 stables and 1,443 horses, mostly in the one- or two-horse shacks of peddlers and express drivers. He reported that 51 percent of the stables were in "bad" condition and only 16 percent were in "good condition." Moreover, Chicago's weak stable statute, which only required annual cleaning, was rarely enforced.[43]

Living near humans was not always good for horses either. Some boys delighted in teasing them. Humans could also start fires, make nighttime noises, spread disease, and dispose of their wastes in an unhealthy manner.

A substantial number of wealthy urbanites maintained their own stables and horses. The 1865 *Report of the Citizens' Committee* in New York City stated that, of the 163 private stables in the then affluent Twelfth Sanitary District (Fourteenth Street, the Bowery, Houston Street, and Sixth Avenue), "very many are neat and most carefully kept."[44] Wealthy New Yorkers who could afford individual stables and carriage houses originally built them behind their homes, but in the post–Civil War period they began placing them some distance from the house to escape their smells and noises. In addition, the stables could be put on side streets where land costs were lower. In 1864–66, a row of thirteen brick-fronted stables was erected on Eighteenth Street in Manhattan, near the fashionable Fifth Avenue residential district. Mostly private stables and commercial liveries occupied this street. They often had restrictive covenants on their mortgages and leases prohibiting their conversion to commercial livery stables or to factories. By the late nineteenth and early twentieth centuries, streets reserved for private and livery stables had become common.[45]

Urban carriage houses could be very elaborate. In *The Private Stable* (1899), James Garland noted that "in many instances the seeds of social ambition are first sown in the stable."[46] In more than 450 pages, he set forth advice and rules for costs, ventilation, drainage, stalls, windows and screens, fodder, water, bedding, types of horses and carriages, bits and other aspects of hardware and harness, and rules for servants. His suggestions followed the etiquette rules of the day, as well as providing comparative prices and advice on the purchasing of horses, carriages, and equipment.

Garland observed that the choice of a stable would be "largely determined by the number of horses and carriages kept and the location of the owner's house." If, he noted, "the owner lives in the fashionable quarter, it is usually impossible to have the stable an adjunct of the house." He warned, however, that if the stable was more than a half mile distant from the house it would suffer from lack of close attention from the owners, and this might possibly lead to deterioration of conditions. On the other hand, the stable had to be located far enough away so that its smells would not prove "objectionable to the occupants of the dwelling."[47]

Narrow lot sizes limited the design of urban carriage houses. The stable was normally divided into two ground-floor spaces—a front section for carriages and a rear section with horse stalls. Quarters for the coachmen or the groom could be found at the front of the second floor, while the rear was usually a hayloft. These buildings usually had large skylights but front windows only. Architects sometimes designed stables of brick and sandstone, with façades designed in a fashionable manner, such as one built in the round arch style, derived from the German *Rundbogenstil*.[48]

Garland estimated that keeping a stable for one horse in the city cost $3,279 to $7,216 a year, mostly labor costs.[49] Boarding a horse cost $10 a week ($520 a year) and removed the risk of bad smells from the home, as well as the need to pay and supervise a servant.[50] Boarding did have risks. Boarding stables increased the danger of disease or injury or mistreatment by hostlers, and dishonest stable keepers might rent horses to others when their owners were not around.

Writing in 1901 about the "Private Stables of Manhattan," Robert Wickliffe Wooley discussed "horse palaces." After describing several of these, he observed that New York's "largest and most pretentious private stable" was that of Louis Stern, on East Eighty-fourth Street. "It should," he noted, "be catalogued as one of the sights of the city." It fronted fifty feet on Eighty-fourth Street, running back for 150 feet. The first floor was divided into three squares 50 by 50 feet each, one of which was devoted to vehicles, the second to fourteen horse stalls, and the third to an arena where the Sterns' children jogged their ponies and where the horses were exercised in bad weather. Overlooking the arena was a balcony for spectators. The second floor had quarters for the coachman, feed rooms, a reception room, a dressing room and gymnasium for men, and a parlor, bedroom, and bathroom for women. The third floor held the quarters for the grooms and footmen. "Everything about it," Wooley assured his readers, "was costly and ornate." Multimillionaire Frank Gould maintained another elaborate private stable on East Fifty-eighth Street that cost $200,000 to erect and was heated in the winter, when riders could exercise in its 50- by 100-foot ring while an automatic organ played in the background.[51]

On Boston's ritzy Beacon Hill, some residents dealt with the stable issue by incorporating the exclusive Beacon Hill Stable Club in 1867. Members paid a membership fee of $1,300 for capital expenses and a monthly fee based on the number of nights that their horses stayed in the stable. The most complete remaining ledger book, for 1912, shows an annual income of $4,218 for twenty-five stalls. The manager spent 28 percent of the budget on food. The biggest expense was labor, 56 percent of the budget, for the manager and eight stable hands, one for every three stalls. The ratio of horses to hostlers was low, less than half that of street railways, suggesting that the horses were comparatively spoiled. The elected board of directors had the power (apparently never exercised) to stop the sale of memberships to people whom they deemed inappropriate. Three women, apparently with full voting privileges, were among the proprietors.[52]

Probably the best-known urban stable was at the White House. Presidents Jefferson, Monroe, Madison, Lincoln, and Grant all added stables on or near the White House grounds. Many of these facilities were inadequate. In 1829, for instance, Congressman Stephan Van Rensselaer, in a report from the Committee on the Public Buildings, noted that "no other than a temporary provision . . . has ever been made for stabling at the President's house." During Jackson's second term, however, the administration constructed a new stable. The *National Intelligencer* described the new building in great detail. The stable was constructed of brick and stone, stuccoed on the outside, and plastered inside, with a supposedly rat-free floor. By 1856, however, the stable was inadequate; a writer in the *United States Magazine* noted that "no respectable drayman in New York with three horses, would have so mean and ill constructed a building—and its location is an outrage upon decency." As with humans, the definition of what constituted adequate housing had changed. A new stable was constructed in 1857 but burned to the ground in 1864. President Ulysses S. Grant built the last presidential stable in 1872. It featured a two-story mansard roof and a frame-and-glass enclosed court. About 1909 President Taft turned the stable into a garage, ordering a White Steamer, two Pierce-Arrows, and a Baker Electric to take the place of the horses and carriages. The structure was finally demolished in 1911.[53]

Most stable hands were hostlers, charged with the day-to-day tasks of feeding, grooming, harnessing, and mucking. Usually, stables employed one hostler for every twelve to sixteen horses. This was an undesirable job, usually filled by recent migrants to the city or unskilled young men entering the workforce for the first time. According to the U.S. Census, Boston had 530 hostlers in 1870, 1,121 in 1880, 1,430 in 1890, and 1,473 in 1900, with disproportionately large numbers of blacks and Irishmen.[54] Hostling usually paid the minimum wage. Like teamsters, hostlers carried bad smells away from their work. The job had long hours with considerable idling time, since horses required cleaning and feeding mostly the start and end of the workday. Horses needed care seven days a week, so many hostlers never had a day off. When they did get Sunday off, humane groups might complain about the poor care that horses got on Sundays.[55] Perhaps because of the idle time, hostlers had a reputation for gambling and drinking. Stables were often the sites of illegal cockfights and of neighborhood fistfights.[56] They were a favorite location for truant schoolboys. One management manual blamed the behavior of stable grooms on their environment, noting that "squalor is not friendly to the maintenance of probity."[57]

James Garland made the most comprehensive complaint, commenting that "coachmen and grooms do not form a class from which angels are exclusively chosen," noting that "drunkenness, brutality, moral obliquity, profanity, laziness, sullenness, and bad manners" were frequent traits. He divided possible stable hands into four categories, starting with "green hands" (men with little or no experience with horses), men who had worked in stables but were out of work (usually "unreliable and of indifferent habits or careless about their work and appearance"), and "sober and industrious" workers fitted to small stables but "incapable of filling more responsible positions." He especially recommended against hiring boys, who inevitably had mischievous instincts, as well as being "arrogant and lazy . . . arrant rascals."[58] This sounds like the usual upper-class Victorian fulminations about servants.

There were positive elements to the job. It was often possible to move up the ladder, becoming a driver or a peddler, both jobs with more prestige and autonomy. A few hostlers wound up owning their own stables or teaming firms. Most probably enjoyed the sociability, even camaraderie, that came with their outcast jobs. The occupation of hostler grew with the horse population. The job, of course, disappeared as horses were phased out.

The horse economy also required large numbers of blacksmiths and harness makers, more specialized and rewarding work. Large stables, such as those operated by the streetcar and cab companies, maintained blacksmith shops. The stable built by the New York & Harlem Railroad in 1876, for instance, contained a blacksmith shop that had eight forges and employed fourteen men. It had the capability of fitting horses with two hundred new shoes daily.[59] Most blacksmiths and harness makers, however, were independent craftspeople.

The Stable Nuisance: Smells, Flies, and Fire

In 1879 the *New York Times* observed that the ubiquitous stable existed even "in the most frequented and fashionable parts of the metropolis" but was especially "undesirable in proximity to private residences." The paper listed the reasons why New Yorkers disliked stables. Stables were "unsightly," and neighbors had to submit to "the rolling in and out of all kinds of vehicles, the stamping of horses in their stalls all night, and the constant yelling of stable-boys and grooms . . . , not what a private family would rank as an . . . inducement to a neighborhood."[60] Other stable noises included the constant whinnying of stabled horses, the bang of the veterinarian's pistol, and the pounding of the blacksmith's hammer.

Smells and manure nuisances were especially objectionable because bad smells were often associated with disease. At least one health survey suggested elevated levels of mortality near stables and among stable hands. Some doctors believed that neighbors risked erysipelas and diphtheria and hostlers risked pneumonia and rheumatism. Stable owners claimed in response that the smells were actually good, encouraging, for example, healthy coughing among tuberculosis victims.[61]

Citizens filed many complaints about stable nuisances and often tried to prevent stable operation in their neighborhoods. In 1877, for instance, the owner of two residences on Fifty-third Street in New York City, Caroline G. Reed, sued the owner of a newly constructed, adjacent livery stable. Reed claimed that when she had purchased her houses she had been assured that "no use" would be made of the adjoining vacant lots "unworthy of the select and fashionable quarter in which they were situated." The stable violated the agreement, and she wanted its operations halted because it was a "nuisance."[62] Such concerns resulted in ordinances separating stables from residences and residential areas.

Another illustrative New York City case occurred in 1897. In that year representatives of the West End Association and the West Side Taxpayers' Association in New York appeared before the Board of Estimate and Apportionment to oppose Street Cleaning Commissioner George E. Waring Jr.'s proposal to erect a department stable in their fashionable neighborhood, located near West Eighty-ninth Street between Columbus and Amsterdam avenues. Protesters complained that the stable would depreciate property values. Waring replied that the stables of the Street Cleaning Department were "conducted so well" that they did not create nuisances. Commissioner Little of the Board of Estimate and Appeal responded that he had inspected two of the department's stables "and had found them anything but the sweet and godly places Col. Waring had described." Members of the Board of Education objected that a new high school was slated for construction across from the proposed stable location and that students should not be subjected to its nuisances. Since a private stable already occupied the site, however, the board decided that a Street Cleaning Department stable was "no more harmful opposite a public school than a private stable."[63]

The smell of manure heaps and the clouds of flies they attracted were some of the most objectionable nuisances stables created. The nuisance was especially bad in summer, when stable keepers tried to "rot" manure before selling it, drawing flies "like sugar-houses." Many manure boxes overflowed, and the manure remained on the ground for long periods. On New York's Liberty Street there was a manure heap seven feet high. New York streets, as shown in pictures from the 1890s, were often covered with layers of manure.[64] Flies were considered particularly objectionable because they were not only nuisances but in fact the carriers of several acute diseases.[65]

Other cities also suffered from manure nuisances. In Pittsburgh in 1894, for instance, the chief sanitary inspector reported 418 manure heaps in violation of a city ordinance. Rats, insects, and flies swarmed over them. But while manure piles were objectionable to those who lived near them, manure sales made an important contribution to the cost of stable upkeep.[66] One businessman in the Boston neighborhood of Jamaica Plain bought an abandoned church on a residential street, planning to convert it to a stable. The building burned to the ground before he could open it, and arson by neighbors who wanted to keep the stable from the neighborhood was suspected.[67] Another excretory problem was urine, which soaked into the stable floorboards. The smell was especially strong in the summer, when it bled out.

Stables posed a major fire threat to life and property. Between 1865 and 1892, the *Times* reported thirty-three stable fires in New York City and vicinity, although, undoubtedly, small stable fires never made it into the newspapers. Those reported ranged in size from the huge Belt Line conflagration to fires in one- or two-horse stables with no deaths but creating a hazard to nearby buildings; it was not uncommon to have a fire in which fifty to one hundred animals perished. In March 1871, for instance, the stables of the Grand Street Railroad Company, located in East Brooklyn, burned. Of the over one hundred horses in the stable, more than fifty were, as the *Times* headlined, "roasted alive."[68] Eighteen months later another Brooklyn streetcar stable fire killed seventy-five horses, and in June 1874 sixty horses died in a fire at the Knickerbocker Ice Company on East 128th Street. The newspapers often noted suspicions that "incendiaries" had set a fire, but these charges were seldom proven.[69]

In October 1881, a very large fire occurred at the New York & Harlem Railroad Company (a Vanderbilt line) streetcar depot and the adjoining Morrell Storage Warehouse at Fourth Avenue and Thirty-second and Thirty-third streets. In 1876 the *Times* had described it as a "model car stable."[70] The press initially reported that the stable's capacity was four hundred horses and that, since many were "in use on the road," only thirty-five had died in the fire. A day later, company officials claimed that 333 horses had actually died, perhaps exaggerating for insurance purposes. Supplies of winter fodder—17,000 bushels of corn, 10,000 bushels of oats, 6,000 bales of hay, and 500 bales of straw—were also destroyed. Another major stable fire occurred in the summer of 1889 at Moses Weil's Empire Livery and Boarding stable on East Eleventh Street near First Avenue. This fire killed 125 horses, as well as destroying 105 vehicles and a large amount of bedding, feed, and harnesses.[71]

Stable fires affected not only their owners and horses but also their neighbors. A week after the 1881 Harlem Railroad fire, the carcasses of the remaining dead horses gave off an "offensive odor," although material that still possessed value, such as rails, had been removed.[72] In the case of the Belt Line fire of May 1887, the fire authorities allowed the fire to burn in order to cremate the horses' remains; in addition, the remaining carcasses were flooded "with a strong soluble disinfectant" to retard smells and possible disease from "miasmas."[73] Still, the carcasses had to be removed and disposed of, and contractors dumped them in the nearby river. Far worse conditions developed as a result of the July 1889 fire at the Empire Livery Stables. Here the July heat caused the carcasses, which were tangled up with other debris, to decompose rapidly, creating intolerable conditions for those living in neighboring tenement houses. "Suffering in Tenements," headlined the *Times,* as the smell made it impossible for residents to open their windows during a stifling heat wave. Again, the Board of Health supervised the application of a liberal amount of disinfectants, thereby reducing nuisances and minimizing, as they believed, the possibility of an epidemic generated by the "stench."[74]

Public regulation reflected popular fears of stables. After 1810 no building could be constructed in Boston without the consent of the city's various legislative bodies. Common Council debates suggest that, de facto, each ward's representative could veto a new stable in his neighborhood. The centrality of health fears in regard to worries about stables was reflected in the fact that the council first referred requests for stable permits to the Board of Health. In 1880 the Common Council authorized 147 stables, almost all small ones in new suburban areas. Concern over fire was demonstrated by the fact that no permit was approved for a wooden stable in the city's older sections. When the South Boston Street Railroad proposed a 150-stall stable in 1871, the largest proposed that year, aldermen insisted that the proposal be kept open for thirty days, advertised in newspapers, and posted in the neighborhood before they would issue final approval.[75]

In the late nineteenth century, most of Boston's rickety old one- or two-horse back alley stables seem to have disappeared. Stringent building regulations following the massive 1872 downtown fire did not allow the construction of new small stables. New stables must have been larger to hold the increased horse population and less flammable. The massive Back Bay landfill of this period, which created a new upper-class neighborhood, made special provisions for horses. Many of the neighborhood's residents owned horses for recreational driving, but they could stable them only on the upper blocks of Newbury Street in three-story brick buildings (wood was allowed for the floors and frame).[76]

New York's Board of Health claimed, probably with exaggeration, that stable regulations were stringent and enforced with more than seven thousand inspections per year. The board's *Annual Reports* provide a wealth of information on stable conditions, especially health-related ones. For example, in 1870 the Board of Health reported that 2,754 of the city's 6,034 stables had responded to a new law and had connected their drains to sewers, a key policy in reducing stable smells from equine urine. Remarkably, the city mandated such connections for stables before it did for human habitations.[77]

Manure removal posed a major regulatory issue for New York. The data for

1870 provides an insight into the problem. Only 253 stables removed manure daily. The board rated slightly more than half of the stables in "good" condition and possessing brick-lined, "air-tight" manure pits, usually holding on average sixty-four cubic feet. While the Board of Health preferred indoor pits to protect human health, stable owners preferred outdoor ones to preserve equine health.[78] Whether indoors or outdoors, manure still had to be removed. In 1870 the city's 38,272 horses produced an estimated 1,146 cartloads of manure daily.[79] Most of this went to the city's fourteen manure dumps, whose owners allowed it to rot, turning it over periodically with pitchforks, to increase its value before sale. The dumps usually but not always were located in "nuisance districts" in the west and east thirties near the rivers and at the city's outskirts, and their size reached epic proportions in the summer, when there was no market for the manure.[80]

Manure storage often aroused fierce opposition from neighborhoods where dumps were located. On November 18, 1884, women from fashionable Beekman Place complained to New York's Board of Health about the twenty-five-foot high (by their reckoning) pile of manure kept in a vacant lot on East Forty-sixth Street. The pile was the responsibility of Tammany-connected manure contractor Martin Kane, who employed one hundred horses to remove manure from stables holding twelve thousand to thirteen thousand horses. Kane stored it, pending sales in the countryside, and if sales were slow or prices low, his inventory grew. The women complained that "sick headaches and nausea have become common on Beekman Hill" and that property values had declined 50 percent because of the odor. The Grand Jury ordered the removal of the manure to abate the nuisance, but the board allowed Kane to re-open his yard with his promise to reduce the size of the manure pile.[81]

Objectionable conditions, however, continued, and a group of more than two hundred women living near the dump signed a petition to the Grand Jury asking it to take action against the nuisance. They also formed the Ladies' Protective Health Association, which proposed to move against not only the manure dumps but also the slaughterhouses in the neighborhood.[82] Kane, the contractor, was brought before the court and charged with maintaining a public nuisance. At his trial, representatives of the association testified about the bad smells emitted by the manure and produced a doctor who validated their health complaints. Kane obtained contrary testimony from a neighborhood dentist, Joseph Conway, who argued that, "while the odors from the manure dump might not be the most pleasing to sensitive nostrils, they were not nauseating and in some cases cured disease." Dentist Conway even claimed that the odor helped his chronically sore throat.[83] The jury, however, was skeptical of these health benefits and convicted

Kane, who had to pay a small fine. But the stench continued, as did neighborhood complaints.[84]

Manure disposal seems the classic urban NIMBY (not in my backyard) problem. According to nineteenth-century health ideas, stables had to remove the manure for the health of both horses and humans, but piling it up someplace for transshipment and ultimate recycling just transferred the same complaints to a different place.

New York conditions slowly improved. By 1890 most of the filthy one-horse stables were gone (health regulations may have increased their costs to an unreasonable level). The number of stables remained around six thousand, while the number of horses had nearly doubled, so the average stable probably held well over twenty horses. The board now displayed more vigor in enforcing the requirement of a covered manure pit and mandated that their contents (more than 450 tons a day) be pressed, baled, and removed daily, not only from the stable but also from the city. The number of inspections remained roughly the same, but the number of citations was down, probably a sign of increased compliance. Stable owners may have recognized that the regulations protected their employees, both equine and human.[85]

In Chicago, Hunter's 1901 report classified 58 percent of the manure boxes as inadequate. Chicago had a weak stable statute, which required only annual cleaning and which usually went unenforced, allowing poor conditions to persist. Hunter found many outside manure heaps, some of which had not been touched in one, even two years. There were twelve overflowing manure boxes in one two-hundred-foot alley near Maxwell Street. A Liberty Street stable had a seven-foot-tall manure pile outside, which reeked so badly that neighbors kept their windows closed all the time, even in the summer. One Chicago physician reported that many small stables opened on alleys where manure heaps and garbage accumulated, "making them as noisome and insanitary as a city refuse heap."[86]

Stables rarely make it into histories of the built environment, although they constituted a substantial part of that environment. Some were multistoried and occupied whole city blocks, housing thousands of horses, their vehicles, their harness and shoeing facilities, their food supplies, and even provisions for the care of sick horses. Thousands of others, smaller and often flimsily built, were scattered throughout the city, occupying back alleys or parts of other structures. As new technologies such as the electric streetcar, the automobile, and the motor truck replaced the horse, many of the stables were torn down or converted to other uses, such as automobile garages. The elimination of many stables, with their horse populations, improved environmental conditions in city neighborhoods, removing health-threatening and smelly piles of manure as well as the other nuisances caused by horses.[87] And their disappearance also undoubtedly reduced a major urban fire hazard.

A few stables remain today in parks, serving their original purpose of providing housing for police or bridle horses, or at racetracks. In some cities former stables even provide residences or house offices or upscale stores. Bus companies often call their storage garages "barns," and sometimes the name is a physical legacy of the days when the label was literally true. Still, most are gone, the victims of back alley cleanup campaigns in the mid-twentieth century or of later slum clearance and redevelopment programs, erasing from the built environment the memory of the horse-powered city.[88]

CHAPTER SIX

Nutrition

Feeding the Urban Horse

As living machines, horses, like any machine, require fuel to function. The "fuel" that urban horses consumed was largely hay and oats. Horses have particular fuel needs because their digestive systems are unusual. They can consume grasses of lower quality than any other ruminant and a lower volume of food than any other large mammal. Moreover, their cecal digestive system allows them to work immediately after eating, while oxen, another draft animal, require hours of rest after eating. The small stomachs and intestines of horses, however, are prone to twisting and blocking (colic). This system requires careful supervision of several feedings a day and a large volume of water. Carbohydrates and nitrogen are more important in their diets than is protein. (They will eat meat, but their system does not require it.) The quantities of feed needed vary with size, workload, breed, and even the weather. L. H. Bailey's *Cyclopedia of American Agriculture* (1908) reported on elaborate experiments with omnibus horses in London and Paris showing that a working horse burned 7,902 calories per 500 pounds and an idle horse burned 4,356 calories per 500 pounds just to sustain bodily functions. This reference work estimated that a walking horse burned 300 calories per 500 pounds and a horse walking up a 10 percent grade burned one-third more energy. To move one ton 20 miles on level road at 2.9 mph in one day required 3,421 calories for a 1,100-pound horse.[1]

In nature, horses are grazers that prefer a diet of grains and grasses. For stable-fed horses, oats and hay have long been the standard and usually adequate diet. *Bailey's Cyclopedia* notes that each pound of hay provided 327 calories and roughage that was essential to keep a horse's delicate intestinal system in order. Oats provided 882 calories per pound. Table 1 shows some of the variations in diet from stable to stable. The variation in daily caloric intake was from 9,000 to nearly 30,000 calories. The reasons for the variations are unclear, but they may have had to do with the size and workload of the animals.

TABLE I
Feeding Practices of Urban Stables

Place	Year	Function	Source	Pounds Fed: Oats	Pounds Fed: Hay	Daily Caloric Intake
New York	1859	Street railways	Herbert	14	10	—
Boston	1897	Carriage	Garland	?	10	—
Generic	1930	"Urban"	Horse Assoc. of Am.	?	8	—
New York	1905	Riding (idle)	*Riding and Driving*	8	6	9,018
Columbus	1919	Municipal	*Eng. & Cont.*	8	14	11,734
New York	1859	Street railways	Eaton	13	8	14,082
Generic	1880	Street railways	*Nat. Livestock Jl.*	16	12	14,112
New York	1905	Riding (in use)	*Riding and Driving*	12	12	14,508
Pittsburgh	1931	Police	*Pitt. and Its Horses*	12	12	14,508
London	1901	Omnibus	Shadwell	17	—	14,994
Cincinnati	1919	Municipal	*Eng. & Cont.*	14	14	16,940
D.C.	1919	Municipal	*Eng. & Cont.*	14	14	16,940
Rochester	1919	Municipal	*Eng. & Cont.*	11	22	16,976
Generic	1884	Street railways	*Am. Vet. Rev.*	20	10	17,640
New York	1900	Draft	Wells Fargo	16	14	18,036
Pittsburgh	1931	Draft	*Pitt. and Its Horses*	18	18	21,644
Generic	1931	Draft	Simms	19	16	21,990
New York	1919	Municipal	*Eng. & Cont.*	26	21	29,979

SOURCES: Edward L. Anderson and Price Collier, *Riding and Driving* (New York: Macmillan Co., 1905), 40; "Cost of Maintaining City Owned Teams," *Engineering and Contracting* (July 2, 1919): 21; Alexander Eaton, *A Practical Treatise on Street or Horse-Power Railways* (Philadelphia, 1859), 96; James Garland, *The Private Stable: Its Establishment, Management, and Appointments* (Boston: Little, Brown, 1899), 391–401; Henry W. Herbert, *Hints to Horse-Keepers* (New York: A. O. Moore, 1859), 135; Horse Association of America, "Grain Surplus Due to Decline in Horses" (1930 leaflet in the National Agricultural Library); R. Kay, "Railroad Horses: Their Selection, Management, Some of Their Diseases and Treatment," *American Veterinary Review* 8 (2–85): 210; *City of Pittsburgh and Its Horses: Facts and Figures Relative to Every Department Using Horses* (Pittsburgh: Pittsburgh Bureau of Horses, 1931), 3; "Feeding Animals," *National Livestock Journal* (1880): 508–10; Roy Shadwell, *Horse Omnibus* (London: P. Woller, 1994), 40; J. A. Simms and J. O. Williams, *Hay Requirements of City Work Horses*, Bulletin 173 (Storrs: Connecticut Agricultural College, May 1931), 8–9.

Meeting Urban Demand

In the early 1930s, the Horse Association of America (HAA) issued several bulletins in which it blamed the agricultural depression of the 1920s on the decline in the numbers of horses and mules in the nation's cities and farms. The loss of the horse and mule market had freed up millions of acres of land that had formerly produced their feed. According to the HAA, each horse and mule engaged in agriculture consumed the product of about two and one-half acres of "fertile corn belt land" per year, while those engaged in nonagricultural work, who worked longer days, consumed the output from four acres of land. When this land was converted to other uses, it glutted the market with a surplus of products and drove prices for agricultural goods down.

The HAA claimed that the decline of the city horse had caused the most devastating effect. Extrapolating from 1900 horse/human population ratios, the association argued that, without the spread of the automobile and motor truck, there would have been approximately 6.5 million urban horses in 1930 rather than the 1.5 million that existed. The HAA estimated that each horse would have consumed about 3 tons of hay and 62½ bushels of oats a year. The extra 5 million horses would have generated a "ready sale and a steady market" for 19.5 million tons of hay and 1,218.75 million bushels of oats produced on 20 million acres of land, assuming steady crop yield. And the agricultural depression would never have occurred![2] Table 1 suggests that the HAA's figures were an understatement—most urban horses were draft animals that consumed more annually than the estimate. There can be little dispute that the amount of land needed to feed urban horses and mules was vast. In short, horses had to eat in order to produce energy, and the food they consumed absorbed the output of large amounts of agricultural land, required massive capital and labor inputs for production and transportation, and necessitated an extensive regional and urban distribution system.

The classic von Thunen rent model of agricultural activity maintains that farmers would produce the highest value and most perishable goods, as well as goods with the greatest bulk, on farms near the urban market in order to keep transport costs low. More distant farmers would grow crops that had less value or keep land for such uses as timber or grasses for grazing. This suggests that farmers nearest the city would specialize in high-bulk goods, such as hay, as well as garden products and dairying, while farmers farther from the city, or even in another region, might produce wheat and oats and, even farther away, raise cattle or horses. Various studies of agricultural regions and their hinterlands have confirmed the general validity of this model, although changes in technology and transportation capabilities caused pattern modification.[3]

Farmers produced hay and oats first to feed their own livestock and then for export. A market mentality on the part of some farmers—that is, an inclination to raise crops for potential profit as well as for local consumption—developed in the late eighteenth and early nineteenth centuries. Within the field of agricultural history, there has been a debate about how widespread this market mentality was compared to an emphasis on the virtues of self-sufficiency or yeomanship. Without entering into this debate on either side, we suggest that farmers who had access to urban markets would produce a surplus for sale if they could avoid excessive risks. The most important variable seems to be the existence of customers willing to pay farmers prices for their goods that would yield them a profit—that is, a price that would cover costs of production and preparation, of transport, and of sale. The growth of cities and towns dependent on horses for transport and power, therefore, provided a major market.[4]

Growing and Harvesting Hay for Horses

America's first cities developed on the East Coast, especially in New England and the Middle Atlantic regions. Colonists extensively farmed the natural saltwater marshes and freshwater meadows near the cities to provide salt hay *(Spartina patens)* for their livestock. In a middle colony, such as New Jersey, between 1697 and 1783, the colonial assembly approved seventy-four statutes called the Meadows Laws, many of which related to the maintenance of salt marshes for the purpose of harvesting the grass.[5] If salt marshes were near a town, once their own needs were satisfied, farmers shipped any surplus to urban markets.

In these areas within reasonable shipping distance of towns, some farmers began to cultivate forage plants. Native saltwater and freshwater meadow hay had less than half the nutritional value of upland or English hay and was, in theory, a less profitable crop. It was, however, accessible to transport by water, far cheaper than by land, given the available infrastructure. Farmers often retained salt hay for their own stock, sending more profitable upland hay to urban markets. In Massachusetts coastal farms during the period from approximately 1770 to 1830, farmers reduced their cultivation of salt marsh and fresh meadow while increasing their acres of "English uplands" sown with imported grasses high in nutrients and very salable in urban markets. By the early nineteenth century, nonnative grasses imported from Europe, such as bluegrass, orchard grass, alfalfa, timothy, and white and red clovers, dominated in the East and provided a major New England crop. English hay sold at twice the price asked for native hay.[6] By the late nineteenth century, state boards of agriculture in Maine, Massachusetts, New Hampshire, and New Jersey were encouraging farmers to consume salt hay for such local purposes as stock feed, animal bedding, and manure mulching but to sell upland hay to urban markets because of its higher nutrient and economic value. Hay was of great importance to the New England economy and, by the early nineteenth century, nonnative grasses were New England's major crops.[7]

Competition from New York farmers had largely forced northeastern farmers out of the commercial production of wheat and corn by the 1840s; increasingly they shifted to such specialized farming as dairying and market gardening for urban markets.[8] Because of its bulk, hay was also best grown near city consumers. By the 1830s, for instance, Essex County, Massachusetts, farmers were cropping their land so as to produce several crops of hay per year. Jobbers would crisscross the landscape in horse carts, purchasing load after load to sell in Boston, Lowell, and Salem.[9] Hay prices increased during the 1830s, and from 1840 to 1860 hay production and the value of the product increased in the six New England states, reflecting the demand from increasing horse populations.[10]

A study of agricultural transformation from 1770 to 1885 in Brookline, Massachusetts, a Boston suburb, illustrates how the town moved from producing largely for itself in the eighteenth century to growing primarily for the Boston market. Initially, farmers raised grain, vegetables, fruit, hay, and livestock, but they began shifting to Indian corn (maize) after the Revolution and then to market gardening and the production of hay. In 1844–45, vegetables, fruit, and hay accounted for 78 percent of the value of the town's agricultural production. Hay output doubled from 1821 to 1840, as did the amount of livestock raised, including horses. As the opportunity for profits from Boston markets grew, Brookline farmers shifted into hay production for the city's sizable horse population, as well as into market gardening.[11] Farmers did not worry about western competition, since hay had a very low value for its volume, making it unlikely to be shipped over long distances. Suffolk County's 369 farms (mostly in the city of Boston) grew 4,139 tons of hay in 1875, making it the second most productive county in the state. Many of these farmers harvested the salt marshes, which surrounded the city, from barges.[12]

While New England was a great hay-growing region, the crop also developed in other regions in the middle of the nineteenth century, especially near urban markets. Both New York City and Philadelphia were large markets for hay. By 1840, the entire region outside Philadelphia had shifted from general farming and wheat growing toward agricultural specialization in a manner that fitted the von Thunen model.[13]

Farmers frequently enlarged the amount of land under cultivation in response to demand. In Brookline, the amount of land being plowed more than doubled between 1771 and 1820; this was also true of New York State between 1855 and 1865, where farmers producing a market surplus invested in much larger areas of improved land than did "non-surplus yeoman" farmers. Much of this improved land came from increases in meadow, which produced quantities of hay that fed both sheep and livestock, with surpluses shipped to the city. Farmers also increased inputs of labor and fertilizer, producing larger hay yields.[14]

Stable manure formed part of an important reciprocal city-country trade. When D. J. Browne wrote one of the earliest American advice manuals on fertilizers, *The American Muck Book: Treating of the Sources, History, and Operations of All the Principal Fertilizers and Manures* (1851), he recommended applying six to eight cartloads to an acre every year.[15] In Kings County, New York (Brooklyn), in the middle and late nineteenth century, farmers purchased increasing amounts of manure, both human and equine, to fertilize their land, increasing their production of hay and oats, as well as garden crops, for the urban market. They viewed Manhattan as a "veritable manure factory" and even imported the fertilizer from as far away as Albany. These farmers applied more equine wastes than did farmers in any other county in the United States. Urban manures eliminated the need for a fallow cycle, so farmers could grow two vegetable crops a year, as well as hay.[16] Brooklyn farmers believed that these urban manures were more valuable than the imported guano fertilizers.[17]

Brooklyn grew into one of the wealthiest agricultural counties in the United States, with hay production peaking at more than seven thousand tons in 1860. Suburban greenhouses emerged in the nineteenth century to provide cities with vegetables and flowers year-round, and greenhouse owners loved the great volumes of urban manure, not just for its fertilizing quality, but also because it heated the greenhouses as it decomposed.[18] Recycling arrangements similar to that in the New York region also existed in other areas close to growing cities such as Baltimore, Boston, and Philadelphia.[19] As with much rural work, manure spreading could be made into fun, and "whiskey-fuelled dung frolics" were a popular rural pastime.[20]

Making hay required expert judgment, especially in selecting a time for cutting and a period for drying. It was an extremely labor-intensive process and included the steps of cutting the grass, spreading it for drying, forming it into cocks, and then loading it on wagons for transport to the barn.[21] Normally, one worker produced somewhere between a third and a half ton of hay per day. The difficulty of harvesting limited the size of the crop grown and effectively capped the number of animals that could be kept in the city.[22]

After 1850, however, American agriculture increasingly replaced human labor with horse-powered machines. The major steps in the haymaking process offering the possibility of replacing labor with capital were cutting, raking, loading, and baling. The first labor-intensive process replaced by a machine was raking. The revolving rake was somewhat more adaptable to rough terrain—that is, it fitted the contours of the surface of the field. Close cutting over rough ground was essential in mowing. With uneven and small fields, cluttered with stumps and stones, the mower was more likely to break down.[23] This was especially true in New England, the largest hay-growing region in the nation until the late nineteenth century. As production-oriented farmers noted in the 1850s, the horse rake was "unquestionably the best labor-saving farm implement of recent invention" and, as contrasted with the mower, "not to be named among the doubtful implements."[24]

Although innovative farmers experimented with horse-pulled rakes of various types in the late eighteenth and early nineteenth centuries, most important was the revolving horse rake, which became commercially available in 1823. In the 1840s and 1850s, inventors and craftsmen patented numerous types of these "revolvers," and also a variety of wheeled riding or sulky rakes. The earliest adoptions came in areas where hay was a cash crop or where sizable dairy herds demanded more forage, such as farms along the Ohio, Hudson, and other river systems allowing the transport of hay to market via water. Between 1850 and 1870, the horse rake—in both the revolving and sulky types—had become a standard implement among market-oriented farmers where hay was an important cash crop.[25]

The harvesting of grass as well as small grains remained demanding and labor intensive. Well into the 1850s, farmers primarily harvested hay by the hand-swung scythe. One goal, difficult to achieve, was to produce a combined machine capable of acting as both a reaper and a mower, the so-called combine, with which a farmer could cut both wheat and hay. Such devices appeared in the 1850s, as did attachments that would convert reapers into mowers. Their cost ranged from $100 to $150, expensive compared to other types of farm machinery, and many farmers questioned their advantage compared to hand methods. In some areas, farm laborers, concerned about job losses, strongly opposed them. By 1860, however, with prices declining, mowers became common farm machines in gathering the grass, especially for farmers with accessibility to markets.[26]

The ability of the farmer to make a profit on hay production depended on more than harvesting and included the steps of loading and storing, bundling, and shipping to market. In the middle and late nineteenth century, technological innovations, also horse-powered, made each of these tasks easier and more cost effective. Before midcentury, eastern farmers often constructed huge barns on their land for hay storage, while outside stacking was more common in the Middle West and on the Great Plains. Building the stack was demanding and dusty, but the invention of the hayfork and the hay stacker in the 1850s and 1860s simplified the tasks and reduced costs significantly.[27] Later farmers would adopt horse-powered conveyor belts to load hay.

Transporting Hay to Market

Transporting hay to market in the nineteenth century was cumbersome and expensive. Without effective means of baling, transportation costs restricted hay to nearby markets, and farmers seldom shipped hay more than twenty or thirty miles.[28] Farmers commonly transported hay to city markets in their own wagons along with other products. Or they might choose to sell "at the barn," where the buyer assumed the baling and hauling costs. Migratory commission merchants then assumed responsibility for transport as well as baling.[29]

In some cases farmers paid teamsters to haul the hay over existing roads. Before railroads were available, teamsters regularly drove large freight wagons on major east-west roads and turnpikes. There were also many part-time rural teamsters, largely farmers and common laborers, who did short-haul odd jobs on a temporary basis. These occasional teamsters invested in horses, oxen, and wagons and would make connections between a farm and a canal terminal, a river port, or a railroad station, the first steps in getting hay to urban markets.[30] As late as 1900, however, one produce shipper noted that a "majority of the shippers are compelled to load and grade hay from the farmers' wagons."[31]

Regardless of the transport method followed, any technique that would reduce the bulk of the hay would reduce costs. In the early eighteenth century, farmers often pressed or twisted hay to save space when shipping it by vessel. New England farmers also loaded pressed hay on ships carrying livestock to foreign markets.[32] A shortage of hay in Chicago drove an early pioneer in hay marketing, C. S. Dole, to buy hay from farmers, "tramp" it in boxes with his feet, and then ship the boxes by water to Chicago.[33]

The major technological advance in hay baling was the invention of the hay press. As early as 1836 agricultural machinery houses were advertising large, horse-powered hay presses that would press five or six tons of hay a day. One ad boasted, "The Press is not likely to get out of order, and is managed without difficulty by the common laborers of the farm." This press sold for $150, however, putting it beyond the budget of most farmers.[34] By the 1850s, numerous manufacturers of agricultural equipment made hay presses and competition drove prices down; in 1860, for instance, the cost of baling was less than one dollar a ton in Ohio.[35]

Some presses were operated by hand, while a horse-powered capstan drove others. In its 1857 catalog, Nourse, Mason & Co., manufacturer of "agricultural and horticultural implements and machines," advertised "Dederick's Patent Parallel Lever Hay Presses" powered by a horse-driven capstan that made a bale in five revolutions. The catalogue claimed that two men and a boy could bale five to nine tons of hay per day. Depending upon size, the Dederick Press sold for $100 to $175. More than a half century later, in 1915, two-horse presses were selling for $150 to $300 and larger presses for more than $500. A two-horse press required four to five men to operate and could bale eight to twelve tons a day.[36] While there were one-horse presses, farmers preferred the two-horse press. Most balers were stationary, and the hay had to be hauled to them and then removed after baling. Since a bale of hay could be handled six to ten times during its trip from farm to market, the number of wires in the bale and their strength determined the integrity of the bale. Often the size of a bale and its ease of handling determined the popularity of a press. By 1912, more than seventy makes of hay presses were on the market.[37]

Only farmers with considerable acreage could afford the purchase price of a press, although farming cooperatives occasionally purchased them. In 1915 the U.S. Department of Agriculture advised small hay growers (less than fifty tons of hay a year) to hire rather than buy a hay press.[38] Hay merchants frequently operated presses at railroad depots and boat landings where they purchased hay from farmers and shipped it. In 1872, a newly patented continuous hay press with a plunger and rammer appeared on the market. Merchants mounted this machine on four wheels and roamed the countryside with mobile presses, purchasing, baling, and pressing the hay on site and then shipping it to market. Increasingly, larger shipping concerns sent "drummers" into the countryside to buy hay and pushed the small shippers out of the market. Contract balers also canvassed rural areas looking for pressing jobs, but they often produced bales containing low-grade hay, since they were paid by tonnage per day.[39]

By the late nineteenth century, greater capital inputs into agricultural production and better fertilizers had increased hay production from approximately one ton per acre to an average of almost a ton and a half, although in some states production was as high as three tons per acre. Railroad cars increased in size and improved presses cut hay bulk, driving freight rates down still further.[40] In 1899, one hay expert estimated that about 27.5 percent of hay harvested was shipped out of the county in which it had been grown. In that year, the five leading cities for hay receipts were New York City, 411,374 tons; Chicago, 197,778 tons; Boston, 184,510 tons; St. Louis, 176,820 tons; and Cincinnati, 102,717 tons.[41]

Between 1879 and 1909 national hay production increased from approximately 35 million to more than 97 million tons. Demand from growing urban horse populations and larger domestic herds of livestock drove the increase.[42] In 1879, farmers grew hay on 10.8 percent of the nation's improved land; by 1889 hay fields had increased to 14.8 percent, and the percentage remained at approximately this level through 1909. Heavily urbanized New England had the highest percentage of land devoted to hay, rising from 32.4 percent in 1879 to 52.3 percent in 1909, an increase of more than 92 percent.[43] New England's share of the nation's hay production, however, dropped from 20.2 percent in 1859 to 4.8 percent in 1909. Suffolk County's (Boston) hay production was half the 1875 level in 1885 and too small for the State Census to count in 1905.[44] The Middle Atlantic states also lost position, dropping from 33.2 percent of all production in 1859 to 11.6 percent in 1909. In contrast, the West North Central states increased from 7.7 percent to 37.3 percent in the same period.[45]

The leading hay-growing states in 1909 were New York, Iowa, and Kansas. New York and Iowa each had approximately 5 million acres in hay and forage crops, while Kansas had over 4 million. Twenty-two states in total each had more than 1 million acres in hay and forage, and sixteen of these states were located in the north. The ten greatest hay-growing states in the nation by tonnage in 1909 were Iowa, New York, Minnesota, Kansas, Nebraska, Wisconsin, Ohio, Illinois, California, and Missouri.

Oats for Horses

Oats were the other critical food for horses, since they are high in fiber and protein and suited to the horse's sensitive digestive system; they also grow in most kinds of soil with adequate drainage. Because oats mature within ninety days, they provide farmers with a quick turnover. Historically, farmers grew oats in conjunction with corn, devoting land to oats early in the spring, before corn, and harvesting it by midsummer, again before corn. The straw of the oat, or oat hay (oats that retained some of their heads), is also very nutritious.[46] From approximately 1840 through 1870, the states of New York, Pennsylvania, and Ohio led in oats production. After 1870, however, just as with wheat and hay, midwestern states like Illinois, Indiana, Iowa, Minnesota, and Wisconsin increasingly outproduced eastern states. Western states had the highest yields, averaging well over thirty bushels per acre. Oats production in western states also increased because of the great expansion of the wheat industry in the middle of the nineteenth century. Wheat production depended upon horse-driven agricultural machinery; therefore, large, wheat-growing regions tended also to be oats-growing areas.[47]

How much oats was grown for urban markets? It is difficult to estimate urban consumption because farmer's fed large amounts of oats to horses and other livestock on the farm, not to mention humans. Before the railroads came, antebellum urban horse populations provided a "profitable outlet for surplus oats that in general could not profitably be hauled any distance."[48] Oats is a very bulky grain, quite sensitive to transportation costs, but the availability of rail transportation made it a more popular crop on prairie farms. The Department of Agriculture estimated that, between 1897 and 1920, the amount of oats shipped out of the county where it was grown ranged from 18 to 31 percent.[49] By the late nineteenth century, the growing and marketing of oats for urban markets, just like those of hay, had been greatly mechanized and modernized.

Until the middle of the nineteenth century, methods of planting seed for grains such as oats, wheat, rye, and barley varied little from centuries-old hand sowing. In the 1840s and 1850s, however, inventors made major improvements in seed or grain drills, greatly reducing the labor of planting. By the mid-1860s, grain farmers in the Mid-Atlantic states and in Ohio had widely adopted seed drills, and farmers in the prairie states soon followed.

Traditionally, farmers harvested small grains such as oats by hand. The sickle was the most common reaping tool until the late eighteenth century, followed by the adoption of the cradle scythe. In the 1830s through 1850s, Obed Hussey and Cyrus Hall McCormick, as well as other inventors, devised successful reapers, and farmers quickly adopted them. Other innovations providing for more efficient grain harvesting, such as headers and binders, followed in the late nineteenth century, reducing the size of the workforce required for the harvest.[50] All the mechanical innovations, however, depended upon further increases in horse power, so living machines on the farm consumed a large proportion of the oats and hay that powered the innovations.

For millennia, grain had been threshed with a flail or under horses' hooves, but in the 1820s and 1830s numerous inexpensive hand- and horse-powered threshing machines appeared on the market. By the 1830s, farmers could buy more than seven hundred varieties of threshing machines, mostly "groundhog" threshers, so called because they were staked to the ground and dug into it when in operation. Treadmills (also called railway horse-powers) and horse-powered sweeps (often called cider-mill horse-powers) drove the groundhog threshers. Small farmers most commonly used horse-operated treadmills, while large farms adopted horse-powered sweeps. Just as itinerant hay merchants or drummers attached to large firms roamed the countryside looking for hay to press, so did thresher entrepreneurs send crews of workers with horse-powered threshers to seek fields with grain to thresh. These crews were especially active in the Midwest and by 1900 had replaced horse power with more efficient steam power to drive threshers that produced both grain and straw.[51]

Farmers hauled their surplus oats directly to market, sold it to local merchants or itinerant commission merchants, or paid teamsters to take the grain to dealers or commission merchants located in canal terminals, river ports, railroad stations, or cities. Up to midcentury, they usually shipped grain, whether oats or wheat or barley, in sacks. The cheapest way to transport sacks of grain was via river or canal. Sacks were fitted to this form of transport because they could be easily loaded in holds and transferred from boats to piers and then onto wagons. The sack also provided a convenient method of buying and selling, with each sack maintaining its ownership character. Thus, the sack-based water transport grain-marketing system retained "the link between grain as physical object and grain as salable commodity."[52]

The railroad changed these patterns, opening up new regions, lowering transportation costs, and causing farmers to increase production of wheat, oats, and other grains. And, in most cases, horse-powered technologies such as the reaper and the thresher produced these increases. To accommodate and store the vast new flow of grain, however, another new technology, the steam-powered (but originally horse-powered) grain elevator was devised. The efficiency of Chicago's grain-processing technology—the grain elevator, with its ability to handle large quantities—rested upon removing grain from its sacks and moving it in a continuous stream by the railroad carload. Before this system could operate at full efficiency, however, an important administrative step was required. This step was the development of standardized grades of inspection, weight, and quality for grain.

Organizing Markets for Hay and Oats

The organization of markets for oats and hay happened at different times. An organized market was one having official inspection, standard methods of weighing and quotations, and supervision by an organization of merchants specializing in the sale of the commodity. At midcentury, boards of trade in various port and riverine cities around the nation, such as Buffalo, Chicago, New York, and Philadelphia, took the lead in setting standards and organizing grain markets. In 1856, for example, the Chicago Board of Trade set standards of quality for categories of wheat, corn, oats, rye, and barley. The New York Produce Exchange followed in 1874. The Chicago standards tended to become the national norm.[53] This made possible the mixing of grains and the development of transactions in receipts for equally graded grain.[54] Thus, the actual grain was separated from its ownership rights. Other cities like Buffalo, Cincinnati, and Pittsburgh also established Flour and Grain Exchanges.

Oats were classified according to color, with standards set for white, red, gray, black, and mixed categories. They were usually marketed like other grains such as wheat but did not attract as much speculative interest and were frequently sold to buyers directly from rail cars. As with hay, commission merchants eventually became the major outlet for oats entering the domestic market. Unlike wheat, oats were seldom exported.[55]

Standards for hay came somewhat later than those for oats because hay remained a locally produced commodity longer than oats. Before the development of the rail networks and long-distance shipping of hay, urban needs for hay, oats, and corn for feeding horses, as has been discussed, were largely filled from farms in the urban hinterland. In New England, for instance, the average hauling distance to market for commodities such as hay, corn, fish, and potatoes in the decades from 1836 to 1855 was about seventeen miles.[56] Farmers would haul hay into cities on their hay wagons, boats, and sleds and sell it at a set location often called the Hay Market, although other commodities besides hay, such as coal and wood, were also sold there.

Municipalities regulated the location and the operations of these markets. At Chicago's hay market on West Randolph Street, for example, traders could sell hay only at designated stands. In Buffalo, hay or straw sales were restricted to the Clinton Market. In Battle Creek, Michigan, the municipal authorities fined sellers of hay or of wood if they tried to sell their products anywhere except on Jackson Street. In Galesburg, Illinois, the municipality allocated space around the public square to teams selling hay, coal, and wood, with each team standing "lengthwise along the places appropriated to each, close behind each other." And, in Detroit, hay sellers had to locate their wagons and sleds "at, or adjoining the hay scales in regular order, one after another," with fines of up to fifty dollars levied for obstruction of the street.[57]

Hay markets were colorful places, filled with farmers' hay wagons and hay purchasers haggling over prices while their horses stood by patiently. City governments carefully monitored weights, and in most cities a weighmaster had to weigh hay offered for sale on an authorized city scale, reflecting the municipality's role as custodian of a fair market. In Chicago, the Common Council appointed "City Weighers" on a biennial basis. The weighers had the obligation to furnish themselves with "proper scales," inspected and tested every three months by the city sealer of weights and measures. Their responsibility was to "weigh any cart, wagon or sled load of hay," as well as bales of hay. The same regulations applied in Battle Creek, Detroit, Galesburg, and Pittsburgh. Weighers or weighmasters in Chicago, as in other cities, were obligated to provide the seller with a certificate, and persons selling or purchasing hay without a certificate were subject to a fine.[58]

As the urban demand for hay increased, local producers were often unable to meet market needs. Hay increasingly became a commodity in interstate commerce, necessitating the development of a more organized market. In 1877, the Chicago Board of Trade set its first standards for hay. The New York Produce Exchange established the Manhattan Hay and Produce Exchange (later the New York Hay and Straw Exchange) in 1883 with "Rules Regulating the Hay and Straw Trade Among Members of the New York Produce Exchange."[59] The entrance of hay into organized markets was somewhat more complicated than that of oats and other grains and involved issues of packaging, shipping, storage, and standards. As hay production increased in response to growing demand, its shipment and sales were plagued with problems of poorly compressed bales, of grading, and of storage that had largely been solved in the transportation, warehousing, and sale of grains. Some cities had their own standards relating to the sale and grading of hay that only applied to that municipality, with "keen competition" between markets.[60]

Sellers often shipped hay in mixed grades and colors, and carloads lacked uniformity. Bale size varied from region to region, and bales were often too light to achieve the minimum twenty thousand pounds of total weight on a railroad car, requiring the shipper to pay extra freight. Because it was so bulky, even after baling, hay remained a low-value commodity, and railroads occasionally embargoed hay shipments in favor of higher value products.[61] The disorganized conditions in the hay "trade" bothered shippers and merchants. Not only price but also such details as baling, weighing, and distribution varied with each transaction.[62] New York City, for instance, although a major center of the hay trade, suffered from chaotic conditions. *The American Grocer* observed in 1882 that the city's "trade" was hampered by "the lack of system in gathering statistical information, the absence of a recognized standard of grade, such as obtains in Chicago, and the need for rules to induce growers to properly cure and prepare the hay for market."[63] It was not until seven years later, however, that the members of the New York Hay and Straw Exchange finally moved to establish standards for grading hay, as well as a method of uniform price quotations. Rather than making grading mandatory for organization members, the association graded only at the request of shippers who paid for it.[64]

Other cities also suffered from hay trade problems. Hay required terminal warehousing facilities rather than grain elevators for storage, but many cities lacked warehouses. In Kansas City, sales often took place directly from the railroad car, with twenty-five or thirty bales stacked outside the car for inspection (called "plugging").[65]

The 1877 standards of the Chicago Board of Trade (CBT) graded different types of timothy, mixed hay, and prime prairie hay. These standards covered mixture, color, curing, and cleanliness.[66] The rules required that all hay sent for inspection had to be graded and marked with the grade when removed from the car bringing it to the city. A "final inspection and plugging" was required "in order to ascertain the sound condition of each bale," and certificates of inspection would give the weight for each bale.[67] After the announcement of the new rules, the CBT began publishing monthly prices for baled hay, as well as occasional discussions of hay production and markets.

In 1885 the CBT announced that the introduction of prairie hay into eastern states had been a "success" and boasted that Chicago was "gaining in importance as a Hay market," handling approximately 100,000 tons in that year.[68] Just as the growth of wheat production in the Middle West had forced eastern farmers to shift to other agricultural products, so did the increase in the growth of western hay force them to shift to other crops and dairying.

The entrance of hay into long-distance transport drove the reorganization of the hay market. Traditionally, the market had many small sellers and buyers, and retailers' orders were limited in both volume and value. Such conditions required "an interconnecting body of independent middlemen," as in other markets with small producers.[69] In these situations, producers sold to jobbers, and the middleman handled the costs of warehousing and shipping and even absorbed credit risks. In many cases, these middlemen and shippers organized in national trade associations. The formation of such an organization in the hay industry occurred in 1893 as a result of the shift from local to long-distance marketing of hay. Under the new conditions, shippers and consignees no longer had personal knowledge of each other nor of the product they were receiving, resulting in increased clashes over grade.[70]

Willis Bullock, publisher of the *Hay Trade Journal*, agitated through the pages of his journal for reform of the hay trade and the formation of a national hay association. In 1895, at a meeting in Cleveland of more than two hundred hay shippers and dealers, the delegates founded the National Hay Association. They established a set of regulations for grading hay similar to those adopted by the CBT and the New York Produce Exchange, as well as rules regarding agreements between buyers and sellers, an arbitration committee, and a committee on legislation and transportation. In 1899, the Chicago Board of Trade *Annual Report* noted that "much improvement is observed in the care and uniformity with which hay is loaded and shipped from the west, which is largely the result of the National Hay Association, which is one of the most important commercial organizations of the country." Eventually, these grading standards spread to most locations in the United States.[71]

The National Hay Association served as a "facilitative organization" that aided middlemen in performing their operations by establishing the "rules of the game," including grading and standards, gathering and disseminating relevant information, and representing hay shippers in opposing increases in railroad rates.[72] Establishing the rules of the game, however, especially compared with other commodities, such as grains, was a difficult task. Grading was uneven, bales varied in size, and localities and sections possessed different customs in hay sorting. Several cities also had established different standards. Farmers, however, were suspicious of middlemen and the grading of their hay and believed that they were better off shipping to markets direct. A poem recited at the Ninth Annual Convention of the Hay Association ridiculed a farmer who thought he could save money by pressing and shipping his hay directly to market but who actually lost money because he did not understand the shipping process and could not meet grading standards. By the first decade of the twentieth century, however, the market was far more organized, having adopted standard grades of timothy, clover, prairie hay, straw, mixed hay, and alfalfa set by the National Hay Association in twenty-four major urban markets.[73]

The Sale of Hay and Oats

The combination of better management, better transportation, and better farming significantly reduced prices. For both hay and oats, prices in 1897 were half of what they had been in 1867. Since feeding was the highest daily expense of urban transportation firms, this undoubtedly played a major role in the extraordinary growth of urban cavalries. It is very uncommon to read accounts of malnourished urban horses after 1873, and they may well have been fed better than their rural counterparts. The price decline might reflect the general deflation of the late nineteenth century, but prices did not go up in the face of increasing demand, stables spent a declining share of their budgets on food, and hay quality improved. Generally, city hay was required to be higher in quality than hay consumed on the farm or sold in the locality where it was grown. Local market hay was sold from producer to consumer and did not have an official grade—rather, it was sold on its "merit" and could be mixed in terms of type. We found only one complaint of an urban horse owner trying to feed trash to horses, an analogous situation.[74] Lower prices meant that horses were better fed. Rising prices for hay and oats after 1900 probably accelerated the decline of the horse, even though the rise was part of a general inflationary spiral.

One hay marketing expert noted that urban buyers preferred higher quality, so "the grade given by the shipper is of utmost importance." However, visual inspection of quality still played a major role. In terms of bulk, hay furnished the largest volume of feed received in the city. In 1894, for instance, the eighty-six members of the Pittsburg [*sic*] Grain and Flour Exchange handled 7,520 carload lots of hay and 3,215 loads of oats, as compared to 3,184 loads of flour and 2,506 loads of wheat.[75]

Hay also had to go from the railhead to the horse's mouth (and, for that matter, beyond). Urban merchants marketed hay in three basic ways. The first was to sell it directly from the railroad car while it sat on the track. Terminal warehouse companies, railroad warehouses, and holding yards also sold hay. The most common method of marketing hay involved the private warehouse system. In the warehouse, hay would be inspected and graded by type and sold in lots of uniform grade. In some cities, such as San Francisco, the hay was sold at auction in a holding yard.[76] These methods of sale contrasted with the earlier, more casual period when local farmers had sold directly from their wagons. In 1900, the *New York Times* observed that the "old-style load of loose hay, which wound its picturesque way through the city, is no longer seen . . . The hay received in the city is all baled."[77]

Owners of only a few horses, or perhaps only one, purchased their hay, oats, and other feed from feedstores, which, if Boston is typical, were near the hay market. These stores frequently handled flour, seeds, potatoes, and produce along with horse feed. When cities grew, so did the need for intracity freight transport by drays, resulting in an increase in the number of feedstores. In 1857, Pittsburgh chronicler George H. Thurston wrote that "the sales of hay, corn, chopped stuffs and such articles of horse and cow feed, consequent upon the great number of drays here, have given rise to a number of establishments called feed stores." Thurston listed nine principal feedstores that handled about half the city's retail trade in hay, oats, and other feed "stuff." The firms employed twenty-seven hands and did about $214,900 worth of business in that year, involving the sale of 800 tons of hay, 175,000 bushels of oats, and 78,000 bushels of corn.[78] By 1876, at which time Pittsburgh's population and area had substantially increased, the city had sixty-four "flour, grain and feed" establishments, and sixteen years later the Steel City had sixty-three retail flour, grain, and feed establishments and eighteen wholesale establishments. The retail businesses were scattered throughout the city, while the wholesale businesses were concentrated in the "strip" district along Liberty Avenue adjacent to the tracks of the Pennsylvania Railroad.[79]

Such patterns of retail and wholesale distribution centers were also present in other cities. Feedstores generally sold hay in bales, although some loose hay could be purchased. Stable keepers usually kept the bales in a hayloft located over the horses' heads, while storing oats in a grain bin.[80]

Equine Consumption

Because of their small stomachs, horses ate and drank five times a day. If a horse worked a full eight- or ten-hour shift, his teamster would have to feed and water him on the streets. Drivers could either carry feed with them or buy small quantities from local livery stables or, more likely, grocery stores. Commonly, the teamster would hold a feedbag over the horse's head, although sometimes he just threw hay on the street. Teamsters usually carried water buckets, which they filled, perhaps at a nearby saloon. Philanthropic groups and city governments began to build water fountains in the 1860s. The American Society for the Prevention of Cruelty to Animals (ASPCA) built the first one in Union Square, New York City, in 1867, and other cities followed. The Philadelphia Fountain Society, formed in February 1869, had built forty-four fountains by the end of 1871. Both the municipality and volunteer groups built fountains in Brooklyn during the late nineteenth century. Presumably the wealthy horse operators who supported the ASPCA and other animal rights groups hoped that these new urban amenities would also keep their teamsters out of saloons. Such troughs could be a source of disease, especially hoof-and-mouth disease, a zoonose for which humans were the reservoir. Wells Fargo banned its drivers from using fountains, and the veterinary profession recommended their removal.[81]

Underfeeding, although probably uncommon, did appear from time to time. There are some indications that owners returned many animals to the country for winter pasturage before 1860. Some horse-powered industries chose rural locations close to urban markets because turning animals out to graze was cheaper than bringing vast quantities of oats and hay to the city.[82] These practices, which probably led to some underfeeding, largely disappeared with better intercity transportation and declining feed prices.

An 1859 stable manual noted that a cart horse could eat only "as much as his owner could afford."[83] Even after the railroad network was created, the combination of an imperfect national transportation system and an imperfect market led to wild fluctuations in grain prices seasonally, annually, and by location, which probably caused some underfeeding. The extreme seasonality of urban economies exacerbated the problem. To give one measure, as late as 1874 one New York City street railway reported February incomes 56 percent lower than those of July.[84] Owners cut back the rations of living machines in the winter, not just because horses worked less but also because revenues declined significantly and food prices rose. An examination of the *R. G. Dun Credit Reports* for fifty-two Boston stables in the nineteenth century showed that, invariably, stables that borrowed money were small livery operations (carriage rentals were a highly seasonal summer business), whose feed suppliers were extending credit, usually in February or March. Some small businesspersons were likely to skimp on oats because they lacked sufficient capital to get through the winter and could not or would not seek credit.[85] Some winter shortages notwithstanding, the general pattern of equine feeding improved greatly over the century.

Since the energy supplied by the living machine depended so much on food balance, nineteenth-century researchers tried to apply the new science of nutrition to horses. Stable owners closely followed changes in scientific knowledge about animal food requirements. Nutritional science had its roots in Lavoisier's discovery of the chemical basis for respiration and metabolism in the 1770s. Herman von Helmholtz, one of the pioneering researchers in the new science of thermodynamics, had noted that the animal body, like the steam engine, was a site for the dissipation of energy and that fatigue and decay represented a kind of entropy.[86] Justus von Liebig, the prominent chemist, compared human nutrition to a furnace in his 1843 work *Animal Chemistry,* and in 1884 his pupil Ludwig Max Rubner first defined the calorie to measure the comparative energy content of food.[87] European agricultural reformers (and the owners of large urban stables, including omnibus companies in both London and Paris) conducted controlled experiments (for Paris, twenty thousand of them!) in food equivalencies, inventing the notion of the calorie in the mid-1860s to provide comparability between different types of food. Their object was not just to define optimal nutritional requirements but also to find the cheapest food that would satisfy draft animals ("least cost rations"). Horse owners sought to substitute other commodities for relatively expensive oats.[88]

Cornmeal, which was often 50 percent cheaper (although prices fluctuated wildly), was a frequent substitute for oats. Stable keepers believed that four pounds of cornmeal could replace three pounds of oats. New York City street railway owners, probably the earliest disciples of the new nutrition, substituted cornmeal for oats around 1860. The *American Railroad Journal,* which was widely read in street railway circles, reported that using corn could save twenty dollars a horse per year. This created a major nutritional problem, however: corn did not contain the niacin in oats, a fact not known before the discovery of vitamins in the early twentieth century. Corn-fed horses retained the shiny coat usually associated with good health and sustained their size but often lacked the energy for heavy urban work. They also had a higher incidence of colic. Equine life expectancy dropped and horses had trouble completing their routes. This played a role in a series of nasty strikes in which transportation firms tried to undermine popular support for teamsters by claiming that they whipped horses excessively. The drivers responded (quite accurately, with modern hindsight) that corporations provided them with malnourished horses and insisted that they get the same amount of work out of them. They felt that they had little choice but to beat the animals in order to save their jobs.[89] After 1885 few owners fed corn to their horses.

The "corn crisis" notwithstanding, long-term benefits flowed from nutritional experimentation.[90] In midcentury, firms began to chop their grain and to mix it with warm water. One contemporary authority suggested that mixing saved five cents a day by reducing the energy horses spent in digestion.[91] Feeding such mashes, standard by 1860, later allowed some mixture of inexpensive corn (or other grains) with oats, without deleterious nutritional consequences. By 1908 nutritionists had worked out precise, money-saving diets based on horse size, weather, and workload. Increasingly, urban horses lived in larger herds, whose owners had the capital resources to follow an appropriate nutritional regime and avoid winter feeding shortages.[92] The decline in seasonal variations in the economy helped this process, since fewer owners faced a slack season, with the temptation to skimp on feed. They also knew that some animals, frequently called "easy keepers," were better (i.e., less costly) feeders, with Percherons being especially valued for that reason.[93]

Proper feeding practices require not just sustaining equine life and work but also avoiding specific diseases, notably colic, an intestinal blockage caused by indigestion or impaction. Colic is worsened by horses' inability to vomit. It is more a symptom than a disease in its own right, often produced by improper feeding, especially of tainted food, but also by overfeeding or insufficient roughage. We suspect that the impure urban food and water supplies of the nineteenth-century city caused as much water-borne illness to horses as humans, with colic the main symptom. Impure food can also produce salmonella. Colic is also the leading and frequently most fatal symptom of several illnesses, especially intestinal ones. Thus, it is not clear whether the high incidence of colic had nutritional or epidemiological causes. The practices of having horses share the same watering trough and crowding them in large stables could only increase the spread of colic-inducing diseases. One veterinarian reported cases where horses drank too much water after feeding, causing the grain to expand in the digestive tract and block it.[94] Others believed that eating too fast had the same consequences. When veterinarians cited colic as a cause of death, it resembled the vague diagnoses of "old age" or "diarrhea" or "consumption" with which nineteenth-century doctors described the often unknown causes of human mortality.

Four specifically nutritional diseases presented problems in the nineteenth century, especially in cities. At least five epidemics of osteoporosis, or "big head" (a painful swelling of the bones caused by insufficient calcium), occurred, all but one before 1860 and all on the East Coast.[95] Founder, a form of lameness blamed on giving cold water to warm horses, was a chronic issue. Feeding mashes of chopped grain and warm water, a recommended practice, probably reduced its occurrence. Heaves (an equine emphysema) came from moldy or diseased feed, possibly spoiled brewer's grain.[96] Reports of heaves seem less common by 1900, a sign of improved food supplies.

The most common form of nutritionally related disease was azoturia, a form of semiparalysis that was sometimes fatal but usually lasted only several days. The specific etiology involved a combination of overfeeding and underworking and a problem with carbohydrate metabolism. Continuing the high-energy diet of a workhorse on Sundays, when almost all animals had the day off, led to outbreaks of azoturia when the animals were harnessed on Monday. The worst azoturia epidemic on record occurred after the great blizzard of 1888, which shut down New York's streets for several days. Owners continued to feed horses their rich work diets. When they sent the animals out again, more than one hundred died from azoturia, and many others displayed the symptoms for several days.[97] This led to major disruptions of street railway service and freight deliveries. Not surprisingly, the disease was most common among sporadically used carriage horses. Absent disease-specific mortality data, it is impossible to measure the precise impact of any of these illnesses.

Horses used as living machines in the city had to be fueled, and the cost of feeding them depended mostly on the economics of transporting hay and oats over distances, thus conforming to the von Thunen rent model of agricultural activity in relation to urban markets. Production costs diminished over time because of technological innovations in raking, mowing, and reaping. Especially critical, however, was the hay press, which dramatically increased the efficiency of baling and lowered the costs of transport. The growth of the nation's rail network made it possible for hay to be grown in more distant regions and shipped to urban markets. The urban hinterland was then freed up for more profitable products, such as garden produce and milk products, as well as for urban growth itself.

Just as with other commodities, the increases in production and in demand caused major market changes. New specialists, middlemen of one sort or another, created a more formal and organized hay market, with the development of standards for grading, the formulation of more uniform methods of buying and selling the product, and the gathering of statistics. By the turn of the twentieth century, the production, transportation, and marketing of food for horses had become much more systematic and nationally organized, although it was still characterized, as were markets for other agricultural commodities, by numerous inefficiencies.

CHAPTER SEVEN

Health

Equine Disease and Mortality

Health and disease were major problems for any species living in the crowded, often unsanitary conditions of nineteenth-century cities. Urban horses experienced mortality patterns roughly consistent with those of urban humans, with the shorter equine life expectancy taken into account. This is hardly surprising, since horses suffer from roughly 200 of the 250 diseases affecting their fellow mammals.[1] The city was not an ideal habitat for either species, so mortality soared as both urbanized in the nineteenth century. Patterns of equine health under the stress of urbanization help in understanding patterns of human health, and we utilize available mortality data for both species from New York City on a monthly and annual basis.

It took time to develop appropriate technology for food transportation, food storage, and water supply for both species. Poorly drained neighborhoods proved excellent breeding places for insects, such as flies and mosquitoes. Polluted water, dust, and tainted food, all common in cities, were vectors for a variety of diseases. Crowded housing/stabling and frequent contact within huge urban herds increased exposure to contagious disease and fomented epizootics; thus, the animal equivalents of human epidemics were commonplace. The veterinary treatment of horses was relatively rudimentary at midcentury for many of the same reasons as for human medicine and public health. Over the course of the nineteenth century, however, the quality of both human and animal medicine improved, and for many of the same reasons. Reducing equine mortality increased the working life and efficiency of horse power technology, just as reducing human mortality increased the working life and efficiency of human workers. Given the dependence of urban economies on horse power, municipalities pursued public policies to protect equine health.

Urban Horse Mortality

The New York Board of Health compiled horse mortality from the records of the New York Rendering Company, which held the contract to remove dead horses from the city. Since dead horses were a valued commodity that could be processed only with that company's expensive equipment, these data are likely to be fairly accurate. New York was atypical because of its size and density. Smaller or less dense cities in other regions may have had somewhat different patterns of mortality, but New York had the best records that we could locate.[2]

Without knowing the precise causes of most equine deaths, it is still possible to speculate on the reasons for some of the high mortality peaks. October and March are the months with the greatest temperature variations in New York. Biometeorological studies of human populations suggest that such variations lead to mortality spikes.[3] Apparently, great weather variations depress immune systems, increasing the likelihood of infections and making them more lethal. Also, most microorganisms can function only in a very narrow range of temperature and humidity—March and October have the greatest range in those two variables. Almost certainly the late fall increase in horse mortality can be attributed to equine influenza, which recurred annually in the 1870s and 1880s. The early spring spike was consistent not only with rapid temperature changes but possibly with problems in food supply, as winter stocks played out and prices increased.

The March mortality increase may also have resulted from high death rates among urban newcomers. Most owners acquired new horses in the spring, to break them in for the period of peak demand. Stable owners reported many deaths among newly arrived horses in the spring, as they adjusted to urban disease regimes. Stable keepers were careful to make sure that such newcomers received more rest and food than veterans. This explanation assumes that human mortality rose in the spring for some other reason, probably respiratory, since this kind of seasonal migration was purely equine.

The years in which New York's horse mortality most exceeded human mortality on a relative basis were 1872, 1886, and 1887. A massive continent-wide epizootic, probably influenza, occurred in 1872. The 1886 spike was very possibly a local epizootic of glanders, a highly infectious lymphatic disease with respiratory symptoms.[4] The evidence for this is fragmentary. That year neighboring Brooklyn's Board of Health quarantined horses coming from Coney Island, a suspected center of the infection, and suspended one vet who concealed a case of glanders.[5] Reports on the disease also appeared with more frequency than usual that year

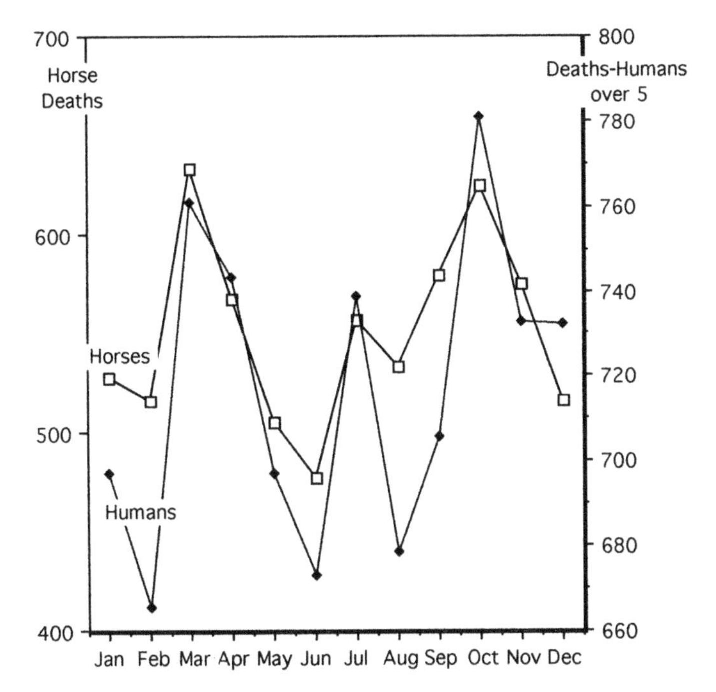

Deaths of Horses and Humans in New York City, 1880–1884. The numbers of deaths (average monthly figures) of humans younger than six years are not included because horses generally came to cities as five-year-olds, and there was high infant mortality among humans. *Open squares,* horses; *solid diamonds,* humans.
Source: Annual Reports, New York City Department of Health.

in the *American Veterinary Review,* which was published in the city. The high 1887 death rate did not result from the horrible Belt Line Street Railway stable fire, since newspaper accounts said that the health authorities dumped the victims in the nearby Hudson River or buried them in the rubble; thus, the carcasses would not have been counted by the New York Rendering Company.[6] The American Society for the Prevention of Cruelty to Animals (ASPCA), however, estimated that it killed forty horses a day during the exceptionally hot summer of 1887, suggesting high mortality from heat stroke.[7] Horses are more likely than humans to work themselves to death in hot weather, especially if a human is driving them, so a heat wave seems a more probable cause of high mortality. This would also explain some of the increase in human mortality.

As we have seen, the ASPCA, acting as an agent of the city's Board of Health, killed roughly one-third of all the horses that died between 1887 and 1897 in New York City.[8] The most common reasons for killing those horses were lameness and

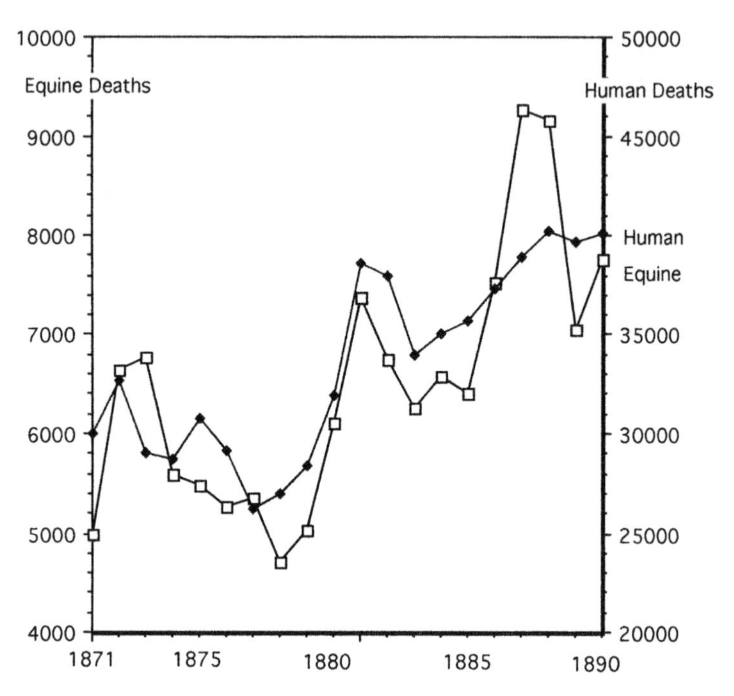

Equine and Human Deaths in New York City, 1871–1890. The numbers of deaths of humans younger than six years are not included because horses generally came to cities as five-year-olds, and there was high infant mortality among humans. *Open squares,* horses; *solid diamonds,* humans.
Source: Annual Reports, New York City Department of Health.

heat stroke. Others suffered from a variety of ailments—most often glanders, already known to be a contagious disease, but also colic. There is no human parallel to this ASPCA function; medical doctors do not kill contagious patients or those too ill to work. Before the ASPCA intervened there were reports of animals too sick or lame to work being turned loose on city streets to die.[9] Veterinarians also killed glandered animals.

Overwork increased urban mortality, and not just in hot weather. Parisian data from 1900 suggest that overwork was a problem, since horses of that city's omnibus/street railway company, who pulled much heavier loads, had double the mortality of cab horses.[10] Overwork was very likely to be a problem in the New World, too. Street railways deliberately sold off their hard-working horses after five years, since mortality rose rapidly thereafter. This suggests that heavy workloads had increased mortality. Overwork could produce both lameness and heat stroke. In some cases American owners intentionally overworked their horses.

The contractors cleaning up San Francisco after the 1906 earthquake knew that they would sell their horses when the job was done, probably at sacrifice prices, so they worked them to death. Replacements were cheap in a rapidly mechanizing society.[11] In much the same way, the contractors who built the New York City subway stabled mules underground, even though they eventually went blind, greatly limiting their resale value. Probably few, if any, of the animals survived.[12]

For both humans and horses, the new industrial city was often a site of violent, accidental death. Fires were especially hazardous for horses, since they were locked in their housing at night, unlike human workers. Industrial accidents also befell horses. For example, in 1884 two ammonia pipes burst in a brewery on a Sunday night and leaked fumes into the stable. By the following morning thirty horses had died and twenty-four "half dead, moaning" ones were led into the street.[13] When a rooftop tank holding 1.5 million gallons of molasses ruptured in Boston in 1919, eleven humans and twenty horses drowned in the street below.[14] Traffic accidents were more commonplace, sometimes merely laming horses, but lameness was often a writ of execution for the urban horse. Accidents caused when brakes failed on a downhill slope, allowing heavy loads to crash into the backs of horses, were often fatal. City life was full of risks.

Zoonoses

In 1870, street railway driver Michael O'Keefe died in delirium and convulsions. His widely publicized death so terrified others that his family buried him immediately and his doctor refused to perform an autopsy. O'Keefe had suffered from glanders, an infection that he had picked up from his horse.[15] Glanders among humans, however, was mercifully rare. In 1884, veterinarian John S. Billings reported three human cases in Philadelphia over the previous twelve years.[16] O'Keefe had probably rubbed a hand with a cut against an open glanders sore in his horse's nose. Apparently, such direct and unusual contact was the only way that glanders could be transported across the species barrier, so human cases were exceedingly unusual. This does, however, raise the issue of zoonoses, infections that spread from species to species. If such crossing were common, it would explain the close relationship of human and equine mortality demonstrated by the New York City data.

Were horse-borne zoonoses increasing human mortality (and vice versa)? Shared diseases include influenza, tuberculosis, pneumonia, bronchitis, pleurisy, some streptococcus infections, diabetes, diphtheria, asthma, anthrax, glanders, thrush (a hoof infection), encephalitis, rabies, hoof-and-mouth disease, tetanus, osteoporosis, rheumatism, emphysema, dysentery, heat stroke, and salmonella. Passage from species to species is rare, however, and some diseases are milder among one species than the other. New York City public health records for humans during the epizootic of 1872 showed no elevation of human mortality from bronchitis, influenza, asthma, or tuberculosis. Nor is there any indication that the deadly human influenza of 1918 affected horses (although it was accompanied by a swine epizootic). It is possible, of course, that other, less deadly influenza outbreaks were zoonoses. Ultimately, it is impossible to identify yesterday's zoonoses with the data available in the historical record.

Tuberculosis, a major killer of humans, seldom affected horses. Anthrax, sporadic among horses, rarely spread to humans and then mostly to animal handlers with a lot of exposure. In theory, the two species could infect each other with rabies, but we found no case of this—both usually got it from dogs. Hoof-and-mouth disease was much more common among cattle than horses. It probably did spread to humans, but only as a mild rash. The diphtheria and typhoid bacilli can breed in horses, but the horses suffer few ill effects. Humans may have infected horses with salmonellosis, but cases would almost certainly have been listed as colic. Humans are the leading reservoir of thrush, and horses are the leading reservoir of tetanus. The only study of pre-1900 epizootics notes three possible zoonoses, a 1699 influenza epidemic, a "throat distemper" in Philadelphia in 1814, and anthrax in Philadelphia in 1838. All three happened before either doctors or veterinarians were very good at diagnosis and must be viewed with skepticism. One nineteenth-century veterinarian claimed he had treated an epizootic of equine typhoid, incorrectly suggesting that it was a zoonose. Humans also blamed horses for one disease, the so-called equine encephalitis, but this was an example of humans confusing correlation with causation. Both species get the disease at the same time, but horses are not the cause. An insect harbored by birds is the vector for both species. In the narrowest epidemiological terms, living in close proximity with the other species had little direct effect on mortality.[17]

Horses and humans, however, do provide vectors that allow the more easier spread of disease within each species. Dust in the air is a vector for many respiratory infections. Ground-up horse manure and soot from the coal-fired boilers of humans made nineteenth-century cities ideal for respiratory infections. Unpaved streets, common in cities, were often dusty, too. Direct human contact with street or stable manure could produce tetanus, the leading reservoir of which is the intestines of horses. Although most large cities had eliminated human (and, for that matter, equine) fecal contamination of water supplies by the end of the century, tainted water supplies were probably a major cause of colic because of the presence of bacteria in the water. Some veterinarians and horse-using firms believed that public water troughs spread glanders as well as diseases of the digestive system. In New York the Board of Health removed them after a 1910 hoof-and-mouth epizootic.[18]

There are few specifics in the nineteenth-century literature about insect-borne diseases, although both humans and horses suffer from them. Poor street construction often left insect-friendly pools of stagnant water. Manure dumps and outdoor privies provided breeding places for flies. Some ticks and other insects took refuge in the hides of horses. There is little in the historical sources about this; the idea of vectors was poorly understood until late in the nineteenth century.

Comparative Data on the Mortality of Horses

There are some comparative mortality data on urban and rural horses in Ohio, U.S. military horses, and horses in the city of Paris. The Ohio data shown in the graph compare the entire state to two urban counties, Cuyahoga (Cleveland) and Hamilton (Cincinnati), and one rural county, Union.[19]

Concerns about reliability notwithstanding, some information can be gleaned from the data. The great variations from place to place and the frequent spikes from year to year represent the pattern one would expect if epizootics were commonplace. Surprisingly, cities had mortality rates much lower than those of rural areas, but this may be an artifact of the imprecise manner in which the state collected the data. There are two possible explanations: Urban horse buyers did not acquire draft animals until they were five years old, so infant mortality could not have been a very large problem in cities. Also, owners sold worn-out urban animals for lighter applications, sometimes in the city, but also for the plow. For example, horses would sometimes go lame on urban streets but still be suited for the lighter work schedules and softer surfaces of the country. Thus, rural counties would have had larger concentrations of both old and young animals, those most susceptible to disease. The presence of many colts and yearlings, therefore, may have given Ohio's Union County, a center of horse breeding, a higher mortality rate.

Cities may have offered better environmental conditions as well, although exposure to epizootic disease was greater there. As we have noted, nutrition improved constantly. City horses probably drank purer water by the end of the century than in earlier years, and big-city stable owners seem to have paid more attention to proper feeding than did many farmers.[20]

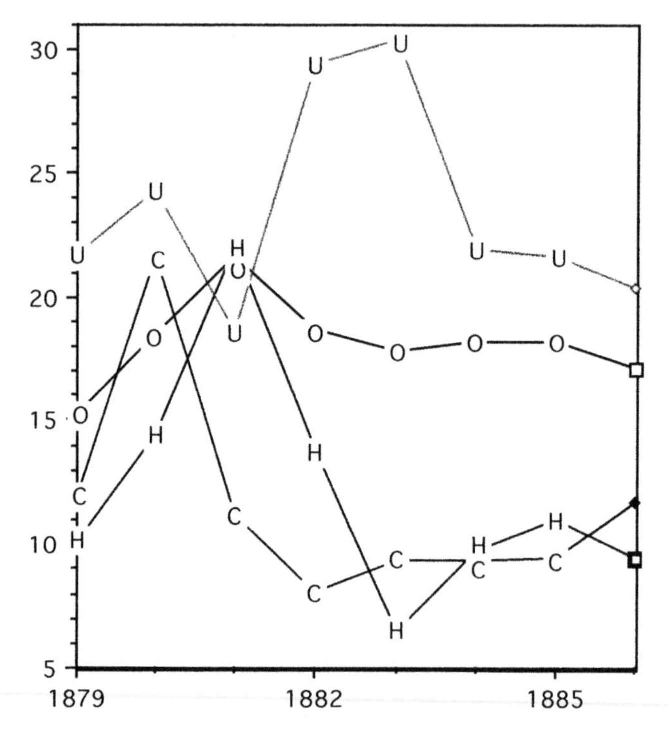

Deaths per Thousand Horses in Ohio, 1879–1886. *O,* entire state; *U,* Union County (rural); *C,* Cuyahoga County (Cleveland); *H,* Hamilton County (Cincinnati).
Source: Annual Reports of the Ohio Board of Agriculture.

Stable-specific mortality data from the Parisian public transport companies, whose careful, detailed horse-keeping records were reported in U.S. periodicals, show a 50 percent decline in mortality for the giant, monopolistic cab company between 1860 and 1890, mostly because of a reduction in respiratory diseases and glanders.[21] Feeding, stabling, and working practices improved during the period. By 1890 glandered horses could be identified fairly well and shot before they spread the disease to other horses. American patterns were probably similar.

American military horses received the same kind of veterinary treatment and endured similar working conditions, although the military was callous, even cavalier, in its treatment of horses. Still, mortality declined between 1865 and 1918. During the First World War, horses serving in the American Third Army in France had a 32.1 percent mortality rate, with less than a third of the deaths related to combat. Horrible as this level of casualties may seem, it was less than half of the Civil War fatality rate for military horses.[22] So the military experience also suggests great improvement in equine mortality.

Veterinary Medicine

Equine health improved during the nineteenth century because of changes in health care. The new veterinary medicine did not cure many diseases, but it did eliminate earlier lethal practices and apply simple public health procedures. The emergence of a laboratory-based veterinary profession paralleled, indeed often anticipated, developments in human medicine.[23]

At the time of the 1850 Census only forty-six Americans called themselves veterinarians. Most of these were probably folk practitioners of farriery (horse podiatry) borrowing a high-falutin' European word. A few had been trained in the fledgling European veterinary schools.[24] The profession was strikingly urban, with twenty of the forty-six practitioners in New York City.[25] An examination of their trade journals suggests that big-city stables provided the largest market for veterinary services. The *American Veterinary Review* described horses as "the most profitable veterinary patients."[26] By the time of the 1890 Census, there were 6,594 veterinarians, almost all in big cities.[27]

Most mid-nineteenth-century health advice about horses grew out of old traditions, perhaps adding some ideas from the latest trends in human medicine. Blacksmiths as well as farriers, repositories of local folk wisdom, often gave health care advice. Some veterinarians were medical doctors hoping for a more lucrative practice than they had with just humans. Benjamin Rush, the most prominent early American physician, had suggested that his students practice on both humans and horses.[28] A few were quacks peddling patent medicines. Although he did not use the word *veterinary*, when Ezra Pater wrote one of the first books about health care for horses in 1794, he captured the confusion of magic, science, and healing: *The Fortune Teller and Experienced Farrier in Two Parts. Part 1: Teaching the Art of Physiognomy and Palmistry, Together with the Significance of Moles, the Interpretation of Dreams &c. Part 2: The Wounds, Sores and Distempers to which Horses are Subject &c. By Ezra Pater, a Jew Doctor in Astronomy and Physic, to which is Added the Wheel of Fortune and How to Use It.* One of Boston's first "veterinarians" advertised herself as a physician, as a veterinarian, and as a spiritualist in the 1873 city directory.[29] The earliest self-proclaimed veterinarians spent much time attacking physicians, blacksmiths, and peddlers of patent medicines.[30]

The primitive nature of traditional veterinary diagnoses can be judged by the symptomatic names assigned diseases. "Glanders" came from one of its symptoms, swollen glands. Influenza was called "snot." Osteoporosis was "big head." Other vaguely described illnesses included burned tongue, staggers, yellow water, and pink eye. The earliest veterinary practitioners were environmentalists, as anti-contagionist as their medical counterparts. For example, one 1852 veterinarian attributed an epizootic to "a subtle poison in the air." As late as 1880, one prominent veterinarian described himself as a "sanitarian." The name had probably lost much of its meaning by then, since he used it in an article advocating the destruction of diseased horses. This was clearly a contagionist policy. Most employed "heroic" remedies borrowed from human medicine, including blistering, heavy purgatives, and bleeding.[31]

Most Americans in midcentury probably treated their horses using advice manuals. British veterinarians William Youatt and Edward Mayhew wrote the best-known works, both of which went through multiple American editions and were available in big-city public libraries. Some of their recommendations were sensible, including avoidance of overfeeding, isolation or execution of glandered horses, and improved stable sanitation. Most folk practitioners had also recommended these measures.

Youatt enjoyed considerable prestige as the editor of the first veterinary journal. He recommended bleeding (sometimes eight to ten quarts), purges using a turpentine-based emetic, and digitalis for respiratory ailments. His suggestions for colic included enemas. Blistering and branding were suggested for other infections.[32] Mayhew advocated lighter bleeding than did Youatt but wanted to treat colic with large doses of opium, ether, laudanum, or belladonna. For respiratory diseases, he suggested liniments involving creosote oil, turpentine, mercury, ammonia, and sulfuric acid. His pharmacopoeia included many poisons, such as lead-, mercury-, or antimony-based medicines.[33]

Probably both Youatt and Mayhew were just adopting the treatments of the heroic human medicine popular in the early-nineteenth-century folk tradition. Their treatments were more likely to kill than cure patients. Bleeding served no therapeutic purpose. Medicines based on heavy metals were poisonous. Purgatives and opiates could exacerbate or even cause colic. Their most important problem was an inability to diagnose properly. Colic was, and to some extent still is, a vague term. It was also hard to distinguish among diseases with respiratory symptoms (early diagnosticians confused glanders and influenza, for example). Both Youatt and Mayhew believed that the latent stage of glanders was a separate disease, which they called farcy, a deadly error. While it is not fair to blame these authors for their diagnostic errors, since they were practicing before the age of the microscope, it is more difficult to excuse the way that their arrogance led them into such murderous treatments. There was some popular skepticism about their doctrines. For example, *Expressman's Monthly* complained, probably accurately, that horses treated "by the books" were more likely to die.[34] Still, these were thought to be the best therapies available.

As horses became more important in the American economy, the demand for better veterinary medicine increased. Agricultural journals led the way, calling for trained veterinarians as early as the 1850s, but the new profession really emerged in big cities. Its practitioners focused mostly on horses, central as they were to city life. Urban stable owners were interested in hiring the graduates of any veterinary school, no matter how scantily trained. The Veterinary College of Philadelphia, an outgrowth of the Philadelphia Medical College, opened first, in 1852, but it lasted barely a decade. Other proprietary schools founded in New York and Boston in the 1850s did not last long either, although more successful proprietary veterinary schools were started in the 1870s. Absent laboratory research or clinical training, they offered little improvement over the nostrums of farriers and blacksmiths.[35]

Graduates of these schools did launch what would become the most important professional group, the United States Veterinary Medical Association, in 1863. This group primarily sought better treatment from the military for "professional" veterinarians. The U.S. Army was skeptical of the new profession (with good reason, since most veterinarians followed the precepts of Youatt and Mayhew). Military officers did not like losing control to professionals. They dismissed one veterinary surgeon after he ordered twenty-six glandered horses destroyed. When the military worried about equine health, it thought primarily about lameness and relied on the advice of blacksmiths. Officers did not worry about a high mortality rate, since they assumed that the supply of remounts was endless.[36]

A series of disastrous epizootics during and just after the Civil War increased the demand for better veterinary science. The first of these was a huge outbreak of glanders during the war. Military horses, like urban ones, were prone to epizootics because concentrating large numbers of horses increased exposure. Epizootics, especially glanders, spread even more rapidly among them because military horses were usually tethered together, not kept in separate stalls. In the first two years of the war, 248,000 Union Army horses died from glanders. The giant remount depot at Giesboro in the District of Columbia sometimes housed over thirty thousand horses. Up to three hundred died there each day, and more than 10 percent died before reaching the front. Giesboro required twenty-six soldiers just to bury dead animals. Horses were not safe from disease at the front, either. Fewer than 10 percent of military horses died in combat. The military sold off infected horses with healthy ones after the war, spreading the disease nationally. They knew better, as evidenced by the fact that cavalry commanders often left glandered horses behind when they retreated, hoping to infect the other side's steeds. The Confederate experience was worse than that of the Union Army, with 75 percent annual equine mortality, little of it in combat. The average life of a Confederate artillery horse was 7.5 months, and one infirmary had an 85 percent mortality rate.[37]

The horses shipped long distances from military depots spread other diseases as well. Shipping fever, likely a streptococcal infection, first appeared in 1866 among newly urbanized horses, many of them formerly in military service. It became a recurring problem with the advent of a national market for draft animals, centered in Chicago, after the war. High mortality among horses first exposed to urban disease regimes after long travel would continue to be an important part of urban equine mortality. By 1893 veterinarians were recommending public health measures to minimize shipping fever, mostly better sanitation and feeding during railroad journeys.[38]

The military glanders epizootic was well studied, notably by former Confederate veterinarian John Page, who wrote an 1876 book to demonstrate that glanders was clearly contagious and invariably fatal and that farcy was merely a latent stage. His postmortem studies also made diagnosis more certain and helped the importation of contagionist scientific ideas from Europe,[39] although glanders was still hard to separate from other diseases. Contagionist ideas led to a public health solution, which minimized glanders outbreaks. Local and state health boards began to mandate the euthanasia of glandered horses, an approach that American veterinarians knew had cut the incidence of the dreaded disease in the French Army from seventy-nine to forty-four per one thousand horses.[40] By 1893, American veterinarians believed that the disease had been checked.[41]

The second great epizootic—an 1865 outbreak of rinderpest in Great Britain that killed five million cattle—was neither American nor equine, but one that sent shivers through animal-owning Americans. British veterinarians were environmentalists, not contagionists. The disaster was preventable but occurred because Britain did not quarantine cattle shipments from the Continent, where the disease first broke out. Americans reacted by allowing the U.S. Department of Agriculture to quarantine or destroy imported animals for the first time. Rinderpest was so evidently contagionist that it increased the spread of contagionist ideas, especially in Britain, but also in the United States. There were similar domestic concerns over Texas tick fever (another cattle disease). In 1871 the Pork Packers Association of Chicago commissioned a report on Texas fever by the prominent, anti-contagionist British veterinarian John Gamgee. The packers feared a northern embargo on shipments from the South, where the disease originated. The northern cattle-growing industry, which feared losing its herds to disease spread by southern cattle passing through en route to Chicago, rejected Gamgee's ideas and began to insist on the appointment of state veterinarians with the power to ban the Texas-raised cattle that carried the disease. Thus, at the official level, both nationally and locally, contagionist ideas became official doctrine.[42] These events had equine implications. State or city veterinarians received the power to kill not only infected cattle but also glandered horses.

The third epizootic confirming contagionist ideas was the Great Epizootic of 1872, which was probably influenza. We surveyed this elsewhere as the primary illustration of urban reliance on the horse, but it also had an important influence on veterinary thought. Both the New York Board of Health and the American Public Health Association printed widely publicized geographical/epidemiological studies of the kind already common with humans. They traced the epizootic step by step from place to place, beginning in York, Ontario, and then moving steadily south and west to Managua, Nicaragua. Similar tracking was done for a milder influenza epidemic in 1881. The contagionists proved their point by noting that stable conditions, weather, and other environmental conditions varied from place to place with no perceptible effect on the disease. Isolated islands like Key West and Prince Edward Island were exempt. The disease spread more rapidly in the East, where cities were closer together and contact between horses was more common, than in the less densely settled West.[43] The lesson was obvious: more quarantines or the destruction of sick animals was the way to go. Urban public health boards then assumed veterinary powers. New York City's Health Board was the first to appoint a veterinarian, naming Dr. Alexander Liutard in 1876. Liutard condemned 151 horses for glanders in his first three years in office.[44]

The new veterinary public health agencies had teeth. For example, in 1878 a New York City health officer spotted a glandered horse being led out of an auction house. He immediately arrested the seller, Edward Garson, and shot the horse. Garson told the court that he believed the horse merely had influenza. The jury, however, credited the testimony of the veterinarian, and Garson received a six-month jail sentence. John S. Billings recommended quarantine for glandered horses in his pioneering 1884 textbook. Even quarantine was tantamount to a death sentence for urban horses, since owners would not long continue to feed horses that could not work. Other examples of the new powers abound. In 1887 the Massachusetts Cattle Commission ordered 192 of the 1,700 Boston street railway horses quarantined. Quarantines offered some headway against glanders. A Washington, D.C., veterinarian credited stringent quarantine rules for reducing glanders 80 percent in five years.[45]

Veterinarians frequently faced the ethical problems common to all professions in a market economy. The *American Veterinary Review* complained that horse dealers were intimidating or bribing vets to secure clean bills of health for horses that they planned to sell, but where no public health authority existed veterinarians could find other ways to enforce quarantine. Several examples from John S. Billings's veterinary public health textbook demonstrate this. A Springfield, Massachusetts, stable owner fired a veterinarian who had diagnosed glanders in his herd and then sold the horses. The vet sought out the purchasers and became an expert witness in their successful lawsuits. Success as an expert witness, it should be noted, is a mark of the public acceptance of a profession. A Philadelphia veterinarian, fired under similar circumstances, had the ASPCA order the glandered horses off the streets, since working them would have been an act of cruelty. Some business owners only learned the hard way. Billings triumphantly reported on a stable owner who fired his vet after he condemned thirty-six horses, only to have three hundred others die, a quarter of his herd.[46]

In the late nineteenth century, a new medical paradigm, the germ theory, appeared, and its first practical successes were with animals. In 1876, the German researcher Robert Koch developed a treatment for anthrax (a major disease of sheep, occasionally found in horses and people), and five years later the French scientist Louis Pasteur solved the mystery of fowl cholera. American veterinary periodicals swiftly picked up the new ideas. As early as 1880, the *American Veterinary Review (AVR)* translated and published Pasteur's important article, "The Connection of the Germ Theory with the Etiology of Some Common Diseases." Dr. Alexandre Liutard, the French-educated editor of the *AVR*, was largely responsible for spreading the new ideas in the United States. When Pasteur developed a treatment for rabies in 1889, the *AVR* greeted it with the headline "Hydrophobia Now, Glanders Next." Pasteur's discovery did little for horses, however, since it was cheaper to shoot than to treat rabid horses.[47]

In the 1890s, laboratory-based research began to provide effective aid for sick equids. In 1891 French veterinarians developed the Mallein test for glanders, which was soon produced in large quantities by the U.S. Department of Agriculture's Bureau of Animal Industry. While not completely foolproof, the test enormously improved diagnosis and, in the long run, it would lead to the elimination of glanders from North America by facilitating the destruction of diseased animals. Some stable owners remained skeptical, preferring patent medicines, and as late as World War I the military often ignored veterinary warnings to kill glandered horses. Better diagnosis led to increased euthanasia. In Connecticut, for example, the number of glandered horses killed by private and public veterinarians increased by a factor of three between 1907 and 1912.[48]

Researchers produced antitoxins for two zoonoses in the 1890s. The tetanus antitoxin, first developed with research on equine subjects in 1894, saved the lives of many humans and probably a few horses, although tetanus was a milder infection for that species.[49] When Robert Koch developed the diphtheria antitoxin in 1894, American health boards discovered that horses, for whom diphtheria was only a mild infection, offered the ideal breeding place for the antitoxin. Equine sera could be injected into humans, providing immunity. The New York City Health Board was the first to set aside a herd to provide antitoxins for both diseases.[50]

These well-publicized new treatments gave veterinarians the power and prestige connected with scientifically based professions in the late nineteenth century. Science lent credibility to their schools. The American Veterinary College, affiliated with Columbia, was the first "scientific" school to compete with the urban proprietary schools. Other urban universities followed Columbia's lead. Even Harvard briefly had a veterinary school. Iowa offered the first program at a land grant college in 1877. The older, less prestigious proprietary schools steadily lost ground, since they offered little laboratory or even clinical training. Other hallmarks of professionalism appeared, including the first professional journal, the *American Veterinary Review,* initially published in 1877. In 1886 New York State began to license veterinarians, a major step toward professional status. Veterinary historian Susan Jones points out that "veterinary leaders used the power of the state as a tool to regulate their profession, enforce their role as animal experts and capture scientific legitimacy."[51]

Why was the germ theory accepted faster in veterinary than human medicine? The fledgling profession needed a scientific base to give it authority more than did the older profession, so the theory fit its needs. In Britain the new ideas encountered resistance from older veterinary practitioners, but this group was smaller and less influential in the United States. Moreover, the move to laboratory-based medicine occurred first with animals, since there were fewer restraints on experimentation. To give a related example, British veterinary science lagged behind that of France and Germany because strong anti-vivisectionist sentiment made research difficult. American anti-vivisection sentiment was relatively weak. One fundraiser for a veterinary school emphasized the importance of comparative medicine, since experiments done on animals could aid human health by avoiding the groups opposed to the modernization of human medicine. While veterinarians had to deal with quacks selling patent medicines, as did human doctors, they never had to deal with opposition based on religious grounds, for example, from spiritualist or Christian Science groups. There were some veterinary medical sects similar to human ones. For example, in 1894 the *American Veterinary Review* published an article by a homeopath advocating herbal remedies. F. B. Carleton, the author of the piece, had attended the Boston University School of Medicine, which preached homeopathic doctrines, after getting his veterinary degree. Articles like this, however, were exceedingly rare.[52]

In 1910, 11,656 men (and not one woman) in the United States called themselves veterinarians.[53] They had the "science," education, organization, and licensing of a prestigious profession. Whether in regulatory roles or private practice, they had enormous power, including one denied to physicians—they could shoot their patients.

The growth of the veterinary profession, however, was slowing already, as the number of the most important kind of patients began to shrink because of mechanization. The *AVR* had seen this threat as early as 1897, when one urban vet noted that some owners were letting horses die that previously would have received treatment. He blamed a price decline in horses on the increase in bicycles and noted that "veterinarians will soon have to seek new fields."[54] After World War I, this urban, equine-centric profession would do just that, becoming pet-oriented in cities and applying their techniques to other animals in rural areas. There is a certain irony in this story. Just as the profession was developing effective solutions to urban health problems, their patients were becoming redundant.

As for the patients, their health did improve, although little headway was made against the two most important health problems of urban horses: colic and lameness. Colic may have been reduced by sanitary measures improving food and water and by the reduction of diseases for which it was a symptom. Owners continued to kill large numbers of lame and therefore economically worthless horses. The living machine undoubtedly remained less reliable than the pure machine, and owners, although sensitive to the health of living beings, still pursued a style of veterinary treatment that emphasized their profits. From this perspective the interest of horse owners clearly lay in exploiting them as machines while living and as raw materials for rendering plants when dead.

CHAPTER EIGHT

The Decline and Persistence of the Urban Horse

The utilization of the urban horse as a living machine declined in the years around 1900, but the speed of its decline and substitution varied from function to function. In some cases, as in the street railway industry, the change from horse-powered to electric-powered transit occurred with great rapidity. In other cases, however, such as certain types of freight delivery, crowd control, and leisure, the transformation was far slower and is still incomplete. A separate set of social, economic, and cultural circumstances in addition to mechanical technology were required to completely unhorse cities. In this chapter we therefore explore some of the areas where utilization of the horse as a living machine persisted and those areas where substitution occurred rapidly. The final result, of course, would be the nearly full substitution of other technologies for the horse.

The Persistence of the Horse as a Stationary Engine

While writers about the horse in the city usually focus on its work as a hauler of streetcars, cabs, and drays, it also played an important and long-lasting role as a source of stationary power. Oliver Evans received the first American patent for a high-pressure steam engine in 1787, seven years after James Watt had begun to produce them in Scotland. All historical writing about the new steam engine, especially in its stationary applications, seems to assume the inevitability of its triumph. Yet the reality was that the adoption of steam as a power source in cities was relatively slow, not really reaching a level of concentration until the 1870s. Eighty years after Evans filed his patent, for instance, the Ohio State Fair gave prizes to 107 different types of horse-powered stationary machines, most with nonfarm as well as farm applications. Urban stationary power appeared to operate primarily in workshops that needed fractional-horsepower. A study of manufacturing in Philadelphia between 1850 and 1880 emphasizes the limited number of firms with steam or water power and the extent to which older manufacturing processes and forms persisted, especially in low-power industries such as wool-carding, bagging, and rope making.[1]

Horses as living machines provided small manufacturers with flexibility because horses could be added or subtracted as needed. In 1860 the American Institute in New York City held a discussion about motive power, assessing the relative costs of horse and steam power. Horse advocates argued that the cost of steam power was high compared to that of horse power: $300 for a one-horsepower engine, 25 cents a day for coal, and $1.50 per day for a skilled operator, compared to lower initial costs of perhaps $30 for the horse, 40 cents a day for food, and the use of lower priced unskilled labor for horses. Attendees actively discussed which was more reliable. Advocates of the steam engine maintained that the horse, unlike the steam engine, had to be fed whether working or not, while horse defenders argued that the horse could do other things besides power machines, such as hauling a wagon to deliver the finished product. The most telling argument for the horse, perhaps, was that it offered low capital costs for entry-level entrepreneurs and therefore played a major role in stimulating economic growth.[2]

Builders and contactors preferred horses, especially in situations calling for limited and inexpensive mechanical power, as well as for hauling. The average contractor had too little capital to invest in expensive, risky steam engines. Even in large-scale construction operations, horses were vital. Little progress was made, for instance, in improving earth-moving equipment throughout the nineteenth century. As late as 1900, the section on "earth-work" in a widely used civil engineering manual focused on horse-drawn equipment. In 1912 the Clyde Iron Works of Duluth, Minnesota, was still advertising several different horse-powered machines for "contractor's use."[3] One ingenious new deployment of horses and, arguably, their first modest application to urban transport was their use as stationary prime movers turning paddle wheels on ferryboats, paddle wheels that had been invented for steam-powered boats but retrofitted to equine power. Horses were to provide the power to propel ferryboats on some water bodies into the twentieth century.[4] Retrofitting from mechanical power sources was a recurring theme of equine technology.

The relative slowness of the transition is striking. Theoretically, the use of horse powers was obsolescent as soon as Watt invented the high-pressure steam engine at the end of the eighteenth century, but they became increasingly common (especially to judge from patent activity) in the 1840s. The American economy was growing so rapidly that the number of horse-operated machines apparently grew at the same time that their proportion of stationary power shrunk. As late as 1895, nearly one thousand horses operated machinery in various urbanized Massachusetts counties, most commonly in food preparation but additionally in quarrying, brick making, distilling, and cement making. Industrial catalogs suggest that some stationary applications of horses lasted to as late as 1920, some forty years after light internal combustion and electrical motors became available. The disappearance took more than a century because horses in many cases were preferable to steam engines for light or portable applications.[5]

Risk Factors Driving the Decline of the Horse

Horse cars, as the first vehicles of railed mass transit in American cities, clearly increased mobility. But, as a new technology, did the public, whether riders or pedestrians, also regard them as a risk to ride? Not necessarily, especially when compared to steam railroads, although the public often objected to disruption by streetcars of normal street functions. Horses were familiar to urbanites, and since horsecar speeds of six to eight miles per hour were relatively slow, the public did not see them as especially hazardous. They accepted accidents, for instance, as a "fact of life."[6]

Over time, however, the presence of horse-drawn street railways (and other applications of horses for transport) significantly increased both perceived and actual risk on city streets. What was new was not horses themselves, but traffic, especially mixed traffic. In 1885, paving engineer Francis V. Greene performed a count on major streets in ten large cities. Just the fact that such a count was taken is a measure of increased concern with traffic. He found that volumes ranged from 7,811 vehicles a week on Broadway in New York to 4,572 on Douglas Street in Omaha. Iconographic evidence also suggests an increase in traffic in the 1880s, with streets becoming filled with equine congestion. New York and Boston both made some streets one way before automobiles were a major element of traffic. Boston's pioneer (1897) subway was built to allow trolleys to avoid the congestion of horse-drawn traffic on Tremont Street. New York faced a bottleneck largely caused by horse-drawn wagons at the entrance to the Brooklyn Bridge and hired F. Van Z. Lane, arguably the nation's first traffic engineer, to design a new entry pattern. Tunnels, ferry terminals, and park entrances were other choke points. Traffic surveys on suburban roads in Boston, Baltimore, and St. Louis between

1909 and 1913, late in the horse era, showed that horse-drawn vehicles were still a majority of the growing volume of travel, and the diurnal traffic jam had already become a conspicuous feature of modern city life.[7]

The increase in traffic led to greater public concern about accidents involving streetcars and other horse-drawn vehicles. In 1885 the *Boston Advertiser* noted that fifty people had been injured and eighteen killed by horse railways in the previous year. "Even the steam railways of the State," commented the newspaper, "hardly make a more fatal showing."[8] Equine traffic of various kinds also disrupted traditional patterns of street use, such as children's play, markets, promenades, and parades common to nineteenth-century urban life. Protests against the living machines reflected an anxiety concerning the intrusion of a new technology, even if drawn by a familiar animal, into public space. Accidents involving horses were not unusual before the horsecar appeared on the scene, but they did not necessarily involve interactions between a quasi-mechanical device—the horse, its attachments, and rails—and the vulnerable human body. By the earliest years of the twentieth century, the mix of horse-drawn traffic, bicycles, cars, and pedestrians on city streets had become quite dangerous. Teamsters were particularly at risk in cities with heavy traffic and in the 1890s stood third in New York statistics concerning the frequency of deaths due to accidents in various professions. Between 1899 and 1907, for instance, horse-related fatalities increased nearly 50 percent.[9]

The street railway industry faced especially harsh problems of accident damage. Lawsuits often followed injuries, with litigation becoming increasingly common as accidents multiplied in the late nineteenth century. Beneath this rise in litigation was "the fundamental assumption that liberty and freedom entailed a right to physical integrity that accident and injury denied."[10] Horsecar firms were especially vulnerable, perhaps because of their deep pockets. One street railway manager complained that they had become "common game for accident hunters, and for 'beats' of all kinds to try their fraudulent practices upon." In 1885, the American Street Railway Association began issuing a monthly bulletin listing suits against member companies and recommended "settlement of all suits that can be reasonably settled."[11]

Horse owners felt other winds of political and legal change besides litigation that made it more difficult to keep the living machine housed and operating in the city. Rickety, cheap wooden stables were no longer an option for well-managed businesses. Neighborhoods were increasingly intolerant of stables because of perceived risks of disease and fire. These concerns were represented in city ordinances that limited locations or required more expensive construction. Manure, which had once been a valuable by-product of street railway stables, now became largely a liability. Not only was the price of manure declining because of competition from guano fertilizers, but stables actually had to pay for removal of what had once been a valued commodity. The law of nuisance also evolved, making "that a nuisance, which was not." Before 1890, manure piles were nuisances only if they could be seen or smelled, but when it was demonstrated that they served as a breeding ground for flies that carried a variety of diseases, the piles became nuisances and health hazards. Health boards required (or enforced old rules) that both manure piles and manure carts be covered, raising the cost of handling manure. Litigation about other stable sounds and smells increased, too. In other words, urban residents were becoming more and more sensitive to the presence of these animals as animals, as the living parts of the transportation system became as controversial as the mechanical.[12]

Horse owners saw other clouds on their economic horizons that acted as a disincentive to the continued use of horses. One was costs, and a continued decline in the price of horses and grain seemed unlikely at a time when the supply of untilled, arable land in the United States was believed to be running out. Another was health—since the influenza epizootic of 1872, which had stopped all street railway operations for several weeks, street railway operators had worried about the possibility of a repeat. The azoturia epizootic that followed the 1888 blizzard in New York City confirmed some of these fears. And a third was fire, as numerous stable fires, most notably the one on New York City's Belt Line Street Railway (described in chapter 5), further demonstrated the fragility of the system and the risks associated with it. Even as municipal regulations regarding fire, health, and nuisances were growing more stringent, humane societies with semilegal power were taking control of the horses belonging to both individuals and firms.[13]

The Disappearance of the Horse from Urban Transit

The fastest decline in any urban use of the living machine involved the shift by transit firm after firm from the horsecar to the electric-powered streetcar. The rapid displacement of horse power by electric power grew out of the complicated economics of the transit industry. Many, perhaps most, street railway owners saw their lines as adjuncts to speculation in suburban real estate. Access to downtown via streetcars guaranteed a subdivision's success, as long as residents could get to offices or shops within thirty to forty-five minutes. By the mid-1880s, in many cities with a population of more than 100,000, suburbs had been built out to their maximum desirable commuting distance. As early as 1882, leaders in the industry were following, even underwriting, experiments with mechanical power that would provide faster speeds than the horse, thus extending commuting range and enlarging the value of suburban landholdings.

The living machine obviously had limits, since improvements in breeding technique seemed to have achieved maximum equine size and speed. While incremental reductions in vehicle weight were still possible, no revolution of the kind created by John Stephenson in the 1830s could be visualized for the future. In large cities, such as New York, Boston, and Pittsburgh, maximum rush hour headways had been reached, and it was physically impossible to add more peak hour service. In 1887, for example, New York's Third Avenue Street Railway reported that it was operating on one-minute headways during the rush hour.[14]

Street railway ridership was also changing in both volume and length of trip. On average, each horse was pulling 27 percent more passengers in 1890 than in 1880 and the average length of each passenger trip was longer. The combination of downtown traffic jams, longer routes, and heavier loads stressed horses. On Boston's Metropolitan Street Railway, the average number of passengers per trip increased from thirty-five in 1880 to forty-eight in 1890. The average number of annual horsecar rides taken by residents of New York, per capita, more than quadrupled between 1860 and 1885, even though steam-powered elevated railroads had taken over traffic on the main north-south streets on Manhattan. In Pittsburgh and Philadelphia during the 1870s and 1880s, ridership more than doubled; other cities had similar patterns. This passenger increase slowed horses, not just because of the added weight, but also because of the added time needed to load and unload riders, and more stops and starts increased lameness. The increase in ridership required more, larger, and healthier living machines as well more commodious and lighter rolling stock.[15]

Ridership even outpaced population growth, since cities were growing at their outskirts, requiring longer commutes. Suburban living in the new detached, balloon-frame homes was becoming the American dream. The new suburbanites depended on transit to reach their homes, and they often defined housing taste in moral terms—suburbs were healthier, safer, and environmentally purer than congested downtowns. Municipal governments reflected the desire for suburbs by pursuing a wide variety of pro-growth policies in taxation, in annexation, and in the provision of services to new subdivisions.[16]

Streetcar firms saw the living machine as limiting their growth and the development of highly profitable new subdivisions. They actively sought alternatives to the horse, especially steam driven, no easy task, since cities put many obstacles in the way of steam power. This research and development operation led first to elevated railroads in New York City in the 1870s, built by entrepreneurs already active in horsecars, but el lines required too much capital investment to provide profitable service on any but the most heavily traveled routes. Only New York built any lines before the coming of electric power. Steam-powered elevated trains also blighted the streets on which they operated, so state courts effectively banned them from even Manhattan after 1883.[17] Thus, they offered no solution. Nobody contemplated subways in the age of steam—there was no way to ventilate the smoke.

The cable car was another early alternative to the living machine. It employed a central steam engine to pull a cable in a conduit under the street, a system first employed in San Francisco in 1872. Cars moved by hooking onto the constantly moving cable and stopped by releasing it and braking. Cable systems polluted only at the central power source, were quiet, and moved faster than horses, since the cable usually moved at about ten miles per hour, but not so fast as to attract regulation. The major drawback was cost—five times higher than the price of a horsecar line—so the systems were deployed only in cities such as Chicago, where ridership demands were extremely high, or Pittsburgh, where terrain was so rugged that street railways had to use small cars or extra teams of horses on hills.[18] In most situations, the living machine trumped these competitors.

Electric trolleys, which received their power through an overhead wire and returned it through the rails, provided the revolutionary system that finally made the living machine obsolescent as a streetcar power source. After many attempts by inventors, the engineer Frank Julian Sprague installed the first technically successful electric streetcar in Richmond in 1887. A year later, Henry Whitney, a Boston land speculator, and Charles Francis Adams Jr., scion of the famous Adams family, proved its commercial viability by building a trolley line to connect their massive suburban landholdings in Brookline with downtown Boston. Reports on the windfall profits that Whitney made on his property—he announced in 1891 that the value of Brookline real estate had increased by 20 million dollars over the previous five years—led to the rapid adoption of trolleys in other cities. Early trolley cars traveled at roughly double the speed of their living predecessors, quadrupling the land available for settlement within a half hour of downtown.[19]

Real estate interests thus became the strongest proponents of mechanization. Inventors had little trouble finding investors for even the most hare-brained mechanization schemes, and the industry reported carefully on each trial attempt. The trolley system spread from city to city with incredible speed. Most big-city horsecar firms had begun conversion within two years after the Boston installation. In 1890, mechanized street railways, primarily the infant electric trolleys but also cable cars, elevated lines, and some light steamers, were already hauling about 20 percent of the urban passenger load. By 1893 almost all big-city traction firms had phased out their horses, and by 1902 97 percent of the nation's streetcar trackage was electrified. In 1902 only 8,902 horses remained on streetcar lines, and they pulled only 6 percent of the riders that horses had carried twelve years earlier. Of the 817 streetcar companies in 1902, 747 used electric power (wholly or in part) and 67 used horse power. Almost half the horse-powered traffic was on one New York line, which evidently lacked the political strength to get the conversion to electricity approved. Most of the remaining horses worked on small lines with fewer than ten animals located in southern or western towns. Clearly, as an article in *Munsey's Magazine* in 1913 observed, the "horse has become unprofitable. He is too costly to buy and too costly to keep."[20]

More remarkably, city governments put almost no obstacles in the way of companies seeking to convert, despite the possible dangers of electric power and high speeds. In several cities urbanites complained about the visual pollution of overhead wires, but most were in a rush to get horses off their streets.[21] The danger of a relatively fast mode of transit on city streets appeared less than in the past. Street railways confined their first electric operations to commercial streets that already had heavy traffic. The speed of the early trolley cars, while faster than horsecars, seemed safely low (before the late 1890s, rarely more than twelve miles an hour), so they appeared a safe mechanical alternate to the horsecar.[22]

The rapidity of this almost complete capital-intensive switchover from horse power to electric power—roughly ten years (1888–1902)—is startling, especially given the slow pace at which stationary engines had converted. Horsecars seemed too old-fashioned for cities that prided themselves on their modernity. Once the owners of street railways saw the possibilities of a low-cost form of mechanical power that was acceptable to the public, they switched immediately. If local firms did not convert, public pressure, in the form of grants to electrified competitors, forced them to do so rapidly. Trolleys allowed increases in route lengths, creating windfall real estate gains for corporate insiders. Owners, riders, the public, and regulators were all anxious to get rid of the horse, not only because of its limitations as a machine but also because the externalities it produced—manure, "typhoid flies," and dead horses—had become intolerable in the modern city.[23] Other horse-pulled vehicles in the city, notably the omnibus and the cab, suffered a similar fate.[24]

The Slow Decline and Long Persistence in Commercial Uses of the Horse

Horse-drawn freight in the city fell into several categories, some of which have been described earlier. Horses did a lot of short-distance hauling around docks and in factory and warehouse districts. They also distributed products such as beer, baked goods, ice, hardware, and coal to local merchants using large, five-ton wagons. Merchants, in turn, often delivered their goods to customers using one-horse wagons. As department stores grew, they developed extensive delivery services based on large stables of horses, and such deliveries might be made at a considerable distance from the store. Express companies, such as American Express and Wells Fargo, provided extensive delivery services and had large fleets of horses and vans. And retail businesses delivered such goods as milk and baked products directly to customers along scheduled routes.

Technical refinements in wagons and horses could not keep up with growing intracity freight demand and increased traffic. By the mid-1890s, cargo rates in big cities were increasing more rapidly than intercity costs, as the efficiency of long-distance steam railway operations greatly improved. As with horsecars ten years earlier, it looked as if horse-pulled freight had reached maximum effectiveness, at least for heavy loads. In Chicago, for instance, during the first decade of the twentieth century, tremendous congestion developed in the Central Business District (the Loop) because of the increased amount of goods transport, which absorbed more of the Loop's street space than any other use. Between 1891 and 1905, the number of teams in Chicago had tripled, and confrontations between teamsters and trolleys, which composed almost half of the Loop's traffic in 1906, were frequent.[25]

Here, too, there was much trial and error, with many failed experiments as part of a very long transition. One sign that the horse was reaching its limit in the 1890s was the huge size of some horse-drawn vehicles. More than a fifth of urban wagons weighed over three tons, and heavier weights required long, hard-to-manage multiple-horse teams. By the 1890s, wagon makers were making steel garbage trucks, tank trucks, and even tree transplanters, capable of carrying up to fourteen tons. All imposed an enormous strain on the living machines that pulled them.[26]

All firms that utilized horses for freight and delivery purposes had to organize their services around "horse pace"—a bundle of practices adapted to horse capabilities. One of these practices was based upon the reality that horses became fatigued and needed rest stops. Much of this rest was taken during the time when various business practices occurred, such as inspection by customers of purchased goods, even including the trying on of clothing, or socializing with and sharing a beer with freight recipients.

These customs had evolved over time, perhaps to meet both horse needs for rest and social needs of drivers. Whatever the cause, the efficiency of horse delivery was often compromised by time spent standing. One study, for instance, showed that, of one hundred "truck hours" for New York horse-drawn cartage deliveries, 29.2 hours were spent in "unproductive work" such as waiting and traffic congestion.[27]

The development of both electric- and gasoline-powered trucks in the 1890s and the early twentieth century seemed to promise the demise of the commercial use of the horse. In the middle 1890s, the *New York Times* editorially heralded "The Passing of the Horse," and the *Expressmen's Monthly* trumpeted, "The day of the horse is doomed."[28] Electric and gasoline trucks had different capabilities, and each operated best in different spheres. The advocates of electric vehicles early targeted the local trucking market, long dominated by the living machine, as a niche opportunity. If retail stores or distributors of products like ice, milk, and baked goods wanted to maintain the shape of their horse-based distribution systems, the electric vehicle seemed to make sense. The electric vehicle had similarities to the operating characteristics of the horse but also had advantages in terms of the distances it could travel, especially without the physical requirement for rest periods and its ability to operate in bad weather. Of course, its battery would still require charging.[29]

In contrast, the sphere occupied by the gasoline-powered truck differed from that of either the horse or the electric vehicle. The most important difference was the suitability of the gasoline truck to long-distance service at a cheaper cost. The gasoline truck could theoretically cover twenty times the area served by a horse team, while the electric truck multiplied the area by a factor of only 3.5. To achieve this increase, however, especially on the part of the gasoline truck, required that the whole service and delivery organization of a firm be reorganized away from "horse-pace" principles. Firms that took this step often reorganized completely on efficiency-based principles. Before these efficiencies could be realized, however, street and road surfaces needed to be upgraded, since intercity and suburban roads were in notoriously bad condition and inner-city cobblestone surfaces that well-suited horses' hooves created serious problems of vibration and maintenance for both electrics and gasoline trucks.[30]

Increasingly in this period of transition from horse-drawn commercial and freight deliveries to the use of electric or gasoline trucks, patterns developed. In 1912 the Electric Vehicle Association funded a study by a Massachusetts Institute of Technology professor who examined the delivery patterns of 107 firms. This study found that, in terms of costs per mile, the horse was less costly for distances up to two miles, electric vehicles were cheapest up to a distance of forty-five miles, and gasoline-powered trucks were cheapest transporting goods for distances over forty-five miles. Full appreciation of their capabilities did not occur until the federal government contracted for the development of the heavy-duty chassis Liberty Truck in 1917 for wartime use.[31] The predictions that motor trucks would drive out the urban horse took a long time for fulfillment, confirming the continued importance of the living machine in the short-haul urban market.

In 1919 an unknown sponsor commissioned a report on "the economic status of the horse" by the Green, Fulton, Cunningham Company, an advertising and consulting firm, which came to similar conclusions. The firm conducted a wide range of interviews throughout the country, trying to determine the future of the horse in the American economy. In regard to urban uses, the survey agreed with the conclusion that the horse made economic sense for short-haul work in the city, but its use for long hauls was "out of the question" because of the truck's speed advantage. The horse, however, possessed a particular advantage for hauls from railroad terminals to downtown "business houses" because of the waits that were often encountered—that is, horses were far less costly "at idle" than were motor trucks, whether electric or gasoline. Other advantages of the horse pointed out by the report included its ability to maneuver in city alleys inaccessible for most motor trucks, the fact that it benefited from cobblestone streets, whereas the motor truck found these surfaces damaging, and its tolerance for route deliveries that involved considerable stopping and starting.[32]

Companies that delivered goods to regular customers whose residences or stores were spread along routes also often found it advantageous to continue to use horses. These included bakeries, coal companies, ice companies, and dairy firms. A number of them actually experimented with motor trucks but found that the living machine could service their customers just as efficiently and at a cheaper cost. Speed, for instance, did not vary much if there was a need to cover a stop-and-start route, but the costs of operation were considerably lower. In addition, the constant starting and stopping involved in route delivery put a large strain on truck motors and transmissions and resulted in high depreciation costs. Finally, horses often developed a knowledge of their routes as good as that of their drivers. A well-trained horse that knew a milkman's or baker's route would go from one customer's house to the next unprompted, freeing the driver to carry the product from the street to the front steps. Living machines were smarter than mechanical motors, so milkmen did not need human helpers, since the intellectual labor of the horse substituted for the intellectual labor of a human.[33]

Thus, while the use of the horse to haul heavy freight and deliver goods in the downtown had largely disappeared from the American city by 1930, living machines could still be found performing other commercial and delivery functions.[34] It would be not until well after World War II that horses would be removed from city streets in terms of light delivery functions, and even then isolated cases of their use continued.[35]

Leisure and the Decline of the Horse

As was true in goods hauling, the pace of displacement of the horse in the field of leisure was variable. The urban horse had allowed the affluent to show their wealth but, as internal combustion and electric cars became commonplace after 1900, the very wealthy switched to the new adventure machines in much the same way that they had periodically switched carriage styles. Central Park, the premier display ground for carriages, allowed cars in 1899, a sure sign of acceptance by the wealthy. In 1905 John Jacob Astor, the leader of Manhattan high society and onetime owner of a string of trotting horses, told the press: "A stable of cars is coming to be recognized as the proper thing for a man of wealth." Before Henry Ford began producing the Model T, cars offered little, if any, advantage in price or reliability over carriages. Elite owners invariably had a chauffeur and perhaps a mechanic, as they once had employed a driver and groom for their carriages and wagons. Thus, the car became the newest way to claim status and wealth, as well as an adventure machine that permitted touring in the countryside. As late as 1914 New York and Boston still maintained speedways for private carriages, but by 1920 the automobile had almost totally supplanted the private carriage in the city, and carriage manufacturers and harness supply houses and dealers were ruefully opining that the carriage horse had become a "thing of the past."[36]

On the other hand, officials involved in horse shows, riding academies, and horse racing reported that the interest of equestrians in high-grade horses, especially riding and saddle horses, was "increasing rather than decreasing." Perceptively, the writers of the report noted that, although the age was becoming more mechanical, the interest in the horse as a fine animal was increasing because "a wonderful horse holds the human interest much more than any mechanical exhibit could hope to do." In short, as the horse was becoming less important as a living machine, used for hauling streetcars and freight, its essence as a magnificent animal was being appreciated more.[37]

The displacement of the horse as a living machine occurred very quickly in some applications but quite slowly in others. The shift from the horsecar to the trolley was probably most rapid, and the persistence of horses in route deliveries was probably the slowest. The horse, however, has retained aspects of its role as a leisure item and also as a sign of social status, although it is also used for policing activities in some cities. It has become mostly a companion animal, even a domestic pet. One can still rent horses to ride on urban bridle paths, although usually at fairly high rates, or take a carriage ride in many cities.[38] Only the wealthy can afford the exurban polo field or thoroughbred ownership. High society still attends the horse show. The racetrack, although in decline, still attracts large numbers of bettors. And police still ride geldings in a few cities like New York, where they are perceived as effective in crowd control.[39] The hard-working living machine, once essential for human prosperity in cities, remains, although no longer essential.

The Horse, the Car, and the City

In the preface we noted the 1881 *New York Times* editorial about the indispensability of the horse in the shaping of the nineteenth-century city.[1] Neither the emerging central business district, nor the first streetcar suburbs, nor such outlying recreational sites as Central Park would have been possible without the living machine. Horses also provided almost all freight movement within cities, and they made vital contributions as power sources (horse powers) for motors in manufacturing establishments. Much of the built environment, including stables, of course, but also streets, residences, and warehouses, was built around the horse's needs. Many of these elements, if not most, are still in place. Horses also served as markers of status for elites, as well as basic sources of livelihood for many others.

The living machine was a technology valued mostly for power production, especially in transportation. It was modified to produce more power, mostly by better breeding and feeding, but also by such accoutrements as improved harness, road surfaces, and vehicles.[2] The needs of the urban horse market also revolutionized American agriculture through the demand for better horse production and the less costly provision and transportation of feed. The living elements of this technology caused problems for contemporaries, but they found ways to turn such issues as manure and mortality into profitable recycling opportunities. Urbanites often found cruelty emotionally troubling and worked hard to reduce it. They also supported the creation of a new profession, veterinary medicine, seeking both to increase the durability of their machines and to treat their living companions more kindly.

As a species, horses could not have survived without human intervention. Humans could not have created the wealth-generating (and, for that matter, leisure) opportunities that came with large cities without horses. As the perceptive 1881 *New York Times* editorial observed, "Deprived of their human servitors, the horses would quickly perish; deprived of their equine servitors, the human population in cities . . . would soon be in straits of distress."

There was a cultural component in the switch to automobiles. The urban horse, rarely a symbol of modernity, at least symbolized human progress and the taming of nature. The automobile, however, as it evolved, came to symbolize modernity, while the horse came in many ways to connote traditional rural life. Increasing awareness of accidents and disease highlighted the fears that city dwellers often held of animals in their midst. At the same time, paradoxically, other urbanites were sentimentalizing the horse, just as they held nostalgia for rural life in general. We measure engine performance by horsepower and truck drivers still call themselves teamsters, both links to an equine past. Modern cityscapes contain a huge variety of animal sculptures for a variety of purposes.[3] The horse remains the subject of morality plays, most recently in the film *Seabiscuit*. Urban sports fans go to arenas to watch Mustangs, Stallions, Mavericks, and Stampeders. Cars display horse-shaped ornaments. In the last ten years, more than forty juvenile books about horses, mostly parables, had a wide enough reputation to be acquired by more than six hundred libraries, including another new edition of *Black Beauty*, which still remains on many reading lists for students at junior high schools.[4]

The reality is, however, that, for all of the horse's critical role as a flexible and evolving technology in the nineteenth-century city, it could not accommodate the requirements of the modern city. Some of the factors involved here are huge increases in freight and passenger traffic and city size and resulting demands for more speed, capacity, and endurance. No amount of breeding or nutritional improvement could alter the fact that the horse was still an animal with limits as a living machine operating in a rapidly growing built environment.

As the horse shaped the nineteenth-century city, so motor vehicles created the twentieth-century city. We can only briefly sketch the implications of machine-powered transportation here, although we have done so in more detail elsewhere.[5] Early expectations for motor vehicles were framed by experience with the horse. Trolleys would run on tracks, like horsecars, automobiles would displace private carriages, mostly serving as status symbols, and trucks would haul goods to and from intercity terminals or carry consumer goods to neighborhoods. All of this happened, but the motor vehicle was still no horse. The car is used far more intensely. For example, in Manhattan, there are six residents for every registered motor vehicle, while at the peak of horse use there was one animal for every twenty-seven people. In other words, motors are used four and a half times as much as horses were.[6]

Modern cities could have grown to their current size without motors. The Meso-American city of Tenochtitlán grew to a population of 400,000 without the aid of machines or the labor of any animals except humans. With few animals other than humans doing the city's hard labor, Baghdad had a population of more than one million in 775, as did Beijing in 1800 and London in 1825. All, however, had extremely high densities.[7]

Hopes were high that cars would relieve traffic jams because of their greater speed and shorter length.[8] The volume of traffic and the appropriation of streets for parking, however, eliminated that prospect. Contemporaries also valued the possibility that motorization might encourage sprawl, then viewed as a desirable social goal, much as adoption of the horse had triggered a major wave of suburbanization.[9] Electric trolleys, it was hoped, would allow more home ownership and socially homogeneous suburbs, both desirable policy goals—indeed, "moral imperatives."[10] Motor vehicles, however, far exceeded these expectations. Residential deconcentration and segregation did take place but on an unimagined scale. The motor truck allowed industrial relocation to the suburbs. Office parks on suburban highways, a post-1945 invention, became the centers of the urban economy. Suburban movie megaplexes and stadia—tellingly, there is a Toyota Field but no longer a Polo Grounds—dominate metropolitan entertainment venues. The motor car, however, also brought downtown's troubles to the suburbs. Problems of traffic congestion have followed radial highways and beltways to the suburbs.

The new (if we can use this adjective for something a century old) transportation mode has its own problems of risk and pollution. While the car may be safer, per vehicle, than the horse, the huge increases in traffic volume, mileage, and speed have increased mortality significantly. To the contemporary eye and nose, nineteenth-century equine pollution seems disgusting and unhealthy. It was, however, mitigated by a reasonably efficient recycling system. Modern air pollution has no such mitigation and is probably far deadlier than its predecessor. Manure is biodegradable; automotive emissions are not. Automotive air pollution requires large volumes of traffic—it was not even documented until Dr. Arie Haagen-Smit discovered the nature and causes of photochemical smog in 1952, hence it did not surface as a major complaint in the early twentieth century after the development of more efficient combustion and emission systems. Moreover, horses burned replenishable biomass fuels, while oil reserves are finite.

Other cultural issues grow out of the conversion. Cars, even cheap ones, are major status items for their owners today, an attitude similar to that of elite carriage owners. Automotive transportation offers privacy, a trait especially valued by women who do not like the leering and groping all too common on city streets or public transportation. Cars can carry packages and bags too heavy for pedestrians or transit riders.

Some things have been lost. The constant presence of living, breathing, defecating, and sometimes dying animals was a constant reminder of nature, even in cities, the most artificial of environments.

Preface

1. "The Horse in Cities," *New York Times* (hereafter cited as *NYT*), July 24, 1881. See also "The Passing of the Horse," *NYT*, Feb. 13, 1896, and E. S. Nadal, "The Future of the Horse," *NYT*, Dec. 24, 1899. On other occasions the *Times* had noted that the horse provided aesthetic satisfaction but also allowed exhibitions of human cruelty on the streets. See *NYT*, Dec. 26, 1869, and Oct. 26, 1883.

2. Lewis Atherton, in his *Main Street on the Middle Border* (Bloomington: University of Indiana Press, 1954), also dealt with some of these same themes but in much less detail. See his ch. 2, "The Horse Is King," 33–64, and ch. 7, "Exit the Horse," 217–42.

Introduction

1. Jared Diamond, *Guns, Germs, and Steel: The Fates of Human Societies* (New York: W. W. Norton, 1997), 76–92, 173; Maxwell Lay, *Ways of the World: A History of the World's Roads and the Vehicles That Used Them* (New Brunswick, N.J.: Rutgers University Press, 1992), 23; Stephen Budiansky, *The Nature of Horses: Exploring Equine Evolution, Intelligence and Behavior* (New York: Free Press, 1997), 40.

2. H. H. Cole and W. N. Garrett, *Animal Agriculture: The Biology, Husbandry and Use of Domestic Animals* (New York: W. H. Freeman & Co., 1980), 195–6, 331–8; Budiansky, *The Nature of Horses*, 80–8, 92–3.

3. Carl W. Gay, *Productive Horse Husbandry* (Philadelphia: J. Lippincott & Co., 1914), 1. Bakewell is quoted in Ernle Rowland, *English Farming: Past and Present* (London: Heineman, 1936), 136; the second quote is from W. J. Gordon, *The Horse World of London* (London: Religious Tract Society, 1893), 16. Also see Jim Mason, *An Unnatural Order: Recovering the Roots of Our Domination of Nature and Each Other* (New York: Simon & Schuster, 1993), 37, and Harriet Ritvo, *The Animal Estate: The English and Other Creatures in the Victorian Age* (Cambridge: Harvard University Press, 1987), 2. Here and elsewhere we have cited British works where widely available in the United States.

4. "For a Horse Power for Propelling Machinery," *Journal of the Franklin Institute* 11 (1833): 36; Louis C. Hunter and Lynwood Bryant, *A History of Industrial Power in the United*

States, 1780–1930, vol. 3, *The Transmission of Power* (Cambridge: MIT Press, 1991), 28–9; Brooke Hindle, *Technology in Early America: Needs and Opportunities for Study* (Chapel Hill: University of North Carolina Press, 1966), 120–1.

5. Eugene S. Ferguson, "The Measurement of the 'Man-Day,'" *Scientific American* 225 (Oct. 1971): 76–97. See also William Youatt, *The Horse; with a Treatise on Draught* (London, 1838), 403–52, and André Guillerme, *Batir la ville: Revolutions industrielles dans les materiaux de construction France—Grande Bretagne (1760–1840)* (Paris: Champ Vallon, 1995), 128–9.

6. Asa Briggs, *The Power of Steam: An Illustrated History of the World's Steam Age* (Chicago: University of Chicago Press, 1982), 58; Eugene S. Ferguson, "The Steam Engine before 1830," in Melvin Kranzberg and Carroll W. Pursell Jr., eds., *Technology in Western Civilization*, vol. 1, *The Emergence of Modern Industrial Society, Earliest Times to 1900* (New York: Oxford University Press, 1967), 245–63.

7. Anson Rabinbach, *The Human Motor: Energy, Fatigue and the Origins of Modernity* (Berkeley and Los Angeles: University of California Press, 1990), 3.

8. Frederick John Jervis-Smith, *Dynamometers* (New York: D. Van Nostrand, 1915), 8–9; Hunter and Bryant, *A History of Industrial Power*, 3:122–4; Richard L. Hills and A. J. Pacey, "The Measurement of Power in Early Steam-driven Textile Mills," *Technology and Culture* 13 (1972): 25–43.

9. Ellwood Morris, "On the Tractive Power of the Horse," *Journal of the Franklin Institute* 1 (Aug. 1839): 28; R. H. Thurston, "The Animal as a Prime Mover: Pt. I. The Human Animal as a Vital Prime Mover and a Thought-Machine; The Energetics of the Vital Machine; Its Transformation; Pt. II. Energy Supplied; Power and Efficiency; Internal Work of the Vital Machine," *Journal of the Franklin Institute* 139 (Jan. & Feb. 1895). See also P. Love, "On the Advantages of Horse Carts and Wagons," *Journal of the British Royal Agricultural Society* 7 (1846): 223.

10. Thurston, "Animal as a Prime Mover," 1–20, 100–21; J. Kenneth Major, *Animal-powered Engines* (London: Batsford, 1978), 12.

11. Augustine Wright, *American Street Railways: Their Construction, Equipment and Maintenance* (Chicago, 1888), 194; "Special Report Concerning Relative Costs of Motive Power for Street Railways," *Report of the Annual Proceedings of the American Street Railway Association* 9 (1890): 70–2.

12. "The Way to Wealth for Our City Is on the Railways within the City," *American Railroad Journal* 26 (May 15, 1853): 296; Augustine Wright, "Amount of Horse-Power Used in Propelling Street Cars," *Van Nostrand's Engineering Magazine* 35 (1886): 37; Kevin James Crisman and Arthur B. Cohn, *When Horses Walked on Water: Horse-powered Ferries in Nineteenth-Century America* (Washington, D.C.: Smithsonian Books, 1998). This innovation was a technology borrowed from steam railroads, just as ferry boats driven by horses had adopted the paddlewheels developed for steamboats.

13. "The Buffalo and Niagara Falls Railroad Co. vs. the City of Buffalo," 50 *Hill* 211–2 (1843).

14. "Street and Suburban Locomotion," *Annual Report of the American Institute* (1863), 482–3. For an example, see "The New Steam Car" (Jan. 28, 1865), 100; *NYT*, Oct. 29, 1872. For an early complaint about horses, see "Complaint of Inhabitants Residing near Cather-

ine Market of Waggoners" (Aug. 24, 1818), New York City Common Council Collection, Common Council Papers, 1670–1831 (1818), Box 66, Folder 1535 (Roll 67), Municipal Archives of the City of New York, Chambers Street, New York City. We thank Jared Day for bringing this complaint to our attention.

15. "Motive Power," *Transactions of the American Institute, Proceedings of the Polytechnic Association* (1860), 539; Clay McShane, *Down the Asphalt Path: The Automobile and the American City* (New York: Columbia University Press, 1995), ch. 5.

16. Rabinbach, *The Human Motor,* 20; Robert Thurston, *The Animal as a Machine and a Prime Mover* (New York: John Wiley & Sons, 1894), 37; "Report on Motive Power," *American Railroad Journal* 57 (Oct. 1883): 272–6.

17. Rabinbach, *The Human Motor,* 20; Thurston, "Animal as a Prime Mover." A Google image search with the keywords Muybridge and horse produced 175 hits containing copies of these photos.

18. Ghislaine Bouchet, *Le Cheval à Paris de 1850 à 1914* (Geneva: Libraire Droz, 1993), 155; "An Expert Visits Our San Francisco Stable," *Wells Fargo Messenger* 4 (Aug. 1916): 198. Henry Stephens, *The Farmer's Guide to Scientific and Practical Agriculture: Detailing the Labors of the Farmer, in All Their Variety, and Adapting Them to the Seasons of the Year as They Successively Occur* (New York: L. Scott & Co., 1853), 157, has the earliest mention of blinders we have seen. No patent was taken out for them before 1868. See U.S. Patent Office, *Patents for Inventions, 1790 to 1873 Inclusive* (Washington, D.C., 1874). The earliest reference in the *Oxford English Dictionary* is also to 1868. Paul Pinkerton Foster, "Helping the Work Horses," *Outing* 53 (May 1919): 68–79, has many photos of horses with straw hats.

19. Reginald S. Timmis, *Modern Horse Management* (Toronto: Cassell & Co., 1921), 43; Gordon, *The Horse World of London;* Wright, *American Street Railways,* 194; Scott Molloy, *Trolley Wars: Streetcar Workers on the Line* (Washington, D.C.: Smithsonian Institution, 1996), 117–9; Jennifer Tann, "Horsepower, 1780–1880," in F. M. L. Thompson, ed., *Horses in European Economic History: A Preliminary Canter* (Reading: British Agricultural History Society, 1983), 26–7; A. L. Stimson, *History of the Express Business Including the Origin of the Railway System in America and the Relation of Both in the Increase of New Settlements and the Prosperity of Cities in the United States* (New York: Baker & Godwin, 1881).

20. "Discussion about Horses—Force of Horses and Men Compared," *Annual Report of the American Institute* (1863), 220–1; Edward Edwards, *History of the St. Louis Fire Department* (St. Louis: Central Publishing Co., 1916), 206; "Street Sweeping Machine," *Annual Report of the American Institute* (1863), 485; Clay McShane, "Gelded Age Boston," *New England Quarterly* 74 (Aug. 2001): 278.

21. The blending of horse and mechanical attachments resembles a cyborg. See Charles Hables Gray, Stephen Mentor, and Heidi J. Figueroa-Sarriera, "Cybrogology: Constructing the Knowledge of Cybernetic Organisms," in Charles Hables Gray, with the assistance of Heidi J. Figueroa-Sarriera and Steven Mentor, eds., *The Cyborg Handbook* (New York: Routledge, 1995).

22. Tim Ingold, "From Trust to Domination: An Alternative History of Human-Animal Relations," and Juliet Clutton-Brock, "The Unnatural World: Behavioural Aspects of Humans and Animals in the Process of Domination," in Aubrey Manning and James Serpell, eds., *Animals and Society: Changing Perspectives* (London: Routledge, 1994), 1–22, 23–36;

Lay, *Ways of the World*, 22; Budiansky, *The Nature of Horses*, 66. Elizabeth Abbott, *A History of Celibacy* (New York: Scribner, 1999), 317, attributes the first recorded human examples to the Persians c. 600 BCE. The North American horse became extinct around 11,000 BCE. The horse-based plains Indians encountered by Europeans in the nineteenth century rode horses descended from those brought to Mexico by the Spanish. These were feral animals whose ancestors, probably for hundreds of generations, had had their reproduction controlled by humans.

23. See Budiansky, *The Nature of Horses*, 80–5.

24. Mason, *An Unnatural Order*, 205; Juliet Clutton-Brock, *Horse Power: A History of the Horse and Donkey in Human Societies* (Cambridge: Harvard University Press, 1992), 112; Harold Barclay, *The Role of the Horse in Man's Culture* (London: J. A. Allen, 1980), 125–40.

25. Frances and Joseph Gies, *Cathedral, Forge and Waterwheel: Technology and Invention in the Middle Ages* (New York: HarperCollins, 1994), 218; Lay, *Ways of the World*, 121–3; A. Everitt, "Country Carriers in the Nineteenth Century," *Journal of Transport History*, New Series, 3 (1975–76): 179–202. The most valuable works on American carriages are Thomas A. Kinney, *Carriage Trade: Making Horse Drawn Vehicles in America* (Baltimore: Johns Hopkins University Press, 2004); William Louis Gannon, "Carriage, Coach and Wagon: The Design and Decoration of American Horse-drawn Vehicles," Ph.D. diss., Iowa State University, 1960; and Museums at Stony Brook, *Nineteenth Century American Carriages: Their Manufacture, Decoration, and Use* (Stony Brook: Museums at Stony Brook, 1987).

26. Jim Mason, *An Unnatural Order*, 205.

27. Ibid., 36–40.

28. Bernard Oreste Unti, "The Quality of Mercy: Organized Animal Protection in the United States, 1866–1930," Ph.D. diss., American University, 2002; Virginia Dejohn Anderson, "Animals into the Wilderness: The Development of Livestock Husbandry in the Seventeenth Century Chesapeake," *William and Mary Quarterly* 52 (Apr. 2002). I. N. Phelps Stokes, *The Iconography of Manhattan Island, 1498–1909* (New York: Arno Press, 1967, reprint of the Robert H. Dodd 1915 ed.), vol. 1, has the best collection of eighteenth-century images that we have seen.

29. As quoted in Donald Worster, *Nature's Economy: A History of Ecological Ideas* (Cambridge: Cambridge University Press, 1985), 94.

30. In *Paradise to Be Regained*, as quoted in Arthur Vernon, *The History and Romance of the Horse* (Boston: Waverly House, 1939), 361.

31. Worster, *Nature's Economy*, 94.

32. Charles Dickens, *Hard Times* (online at www.mtholyoke.edu/acad/intrel/hardtime.htm, visited Jan. 31, 2005).

33. "Animals in Fiction," in Victor Watson, ed., *The Cambridge Guide to Children's Books in English* (New York: Cambridge, 2001), 32–7, is an excellent introduction to the literature.

34. Susan Jones, "Animal Value, Veterinary Medicine, and the Domestic Animal Economy in the United States, 1890–1930," Ph.D. diss., University of Pennsylvania, 1997, 16 [published as *Animal Value: Veterinary Medicine and Domestic Creatures in Modern America* (Baltimore: Johns Hopkins University Press, 2003].

35. Nicholas Russell, *Like Engend'ring Like: Heredity and Animal Breeding in Early Modern England* (Cambridge: Cambridge University Press, 1986), 117.

36. Robert Leslie Jones, "The Horse and Mule Industry in Ohio to 1865," *Mississippi Valley Historical Review* (Mar. 1933): 61–88; J. H. Sanders, *Horse-Breeding; Being the General Principles of Heredity Applied to the Business of Breeding Horses, with Instructions for the Management of Stallions, Brood Mares and Young Foals, and Selection of Breeding Stock* (Chicago: J. H. Sanders Publ. Co., 1885), 94; A. A. Holcombe, "Inspection of Horses and Mules for Army Service," *American Veterinary Review* 5 (Feb. 1882): 465; L. Saltonsall, "The Percheron Horse," *Massachusetts Board of Agriculture Annual Report* 12 (1864): 232–4; John E. Russell, "Horses," *Massachusetts Board of Agriculture Annual Report* 30 (1882): 295–304; *Annual Report of the State Board of Agriculture of Ohio* (1861), xvii.

37. Richard Moore-Colyer, "The Trade in British Heavy Horses with the United States and Canada, ca. 1850–1920," unpublished 1994 manuscript in Clay McShane's possession. For example, see John H. Klippert, "Report of an Agricultural Tour in Europe," *Annual Report of the State Board of Agriculture of Ohio* (1867), 173–367; J. H. Klippert, "History of the Percheron Horse," *Annual Report of the State Board of Agriculture of Ohio* (1873), 104–21. See also Robert West Howard, *The Horse in America* (Chicago: Follett, 1965), 176.

38. J. H. Sanders, "Draft Horses of France," *Annual Report of the State Board of Agriculture of Ohio*, 576–85; Howard, *The Horse in America*, 179; G. Arthur Bell, "Breeds of Draft Horses," *USDA Farmer's Bulletin* 619 (Nov. 16, 1914), 7; Keith Chivers, *The Shire Horse: A History of the Breed, the Society, and the Men* (London: J. A. Allen, 1976), 174.

39. Howard, *The Horse in America*, 179; Bell, "Breeds of Draft Horses," 7; John Simmons, *The American Pocket Farrier* (Philadelphia, 1825), 19; Chivers, *The Shire Horse*, 299; Henry W. Herbert, *Hints to Horse-Keepers* (New York: A. O. Moore & Co., 1859), 51; Russell, "Horses," 301, 304; Jones, "Horse and Mule Industry in Ohio," 78; Alvin Howard Sanders, *A History of the Percheron Horse* (Chicago, 1917), 73, 222; Bouchet, *Le Cheval à Paris*, 43; *Rider and Driver* (Apr. 15, 1893): 6.

40. G. L. Carson, *Studies in Horse Breeding* (Norfolk, Neb., 1910), 206.

41. Otto Mayr, *The Growth of Biological Thought* (Cambridge: Harvard University Press, 1982).

42. Sanders, *Horse-Breeding*, 66–76.

43. Ibid., 51. His associate J. H. S. Johnstone favored devices to limit masturbation by stallions (the restraints looked like the collars placed on dogs to keep them from scratching their ears, except in a more anatomically appropriate position) "as long as they weren't too cruel." See J. H. S. Johnstone, *The Horse Book: A Practical Treatise on the American Horse Breeding Industry as Allied to the Farm* (Chicago: Sanders Publ. Co., 1908), 56.

44. "State Fair," *Annual Report of the State Board of Agriculture of Ohio* (1878), n.p.; Edward L. Anderson and Price Collier, *Riding and Driving* (New York: Macmillan Co., 1905), 240; William G. Hill, "Why Aren't Horses Faster?" *Nature* 332 (Apr. 21, 1988): 678; "Percheron-Normans for Japan," *National Livestock Journal* 8 (June 1877): 23; *Army Horses: Remount Systems Abroad, Improvements Suggested at Home* (Washington, D.C., 1908), 43; Harold Sessions, *Two Years with the Remount Commissions* (London: Chapman & Hall, 1903), 129–31; *Livestock Journal*, May 12, 1898.

45. William Harding Carter, *Horses of the World* (Washington, D.C.: National Geographic Society, 1923), 13, has a quantitative survey of horses by breed in the United States, but it only counted registered breed members in a country where registration was ineffec-

tive. He reported: Percherons, 70,613; Belgians, 10,238; Hackneys (a kind of carriage horse), 5,826; Shires, 5,617; and Clydesdales, 4,248. See also G. L. Carson, *Studies in Horse Breeding* (Norfolk, Neb., 1910), 208; Moore-Colyer, "Trade in British Heavy Horses," 11, 24–7; Herman Bidell, L. C. Douglas, Thomas Dykes, George Fleming, Archibald MacNeilage, Gilbert Murray, and W. R. Trotter, *Heavy Horses: Breeding and Management* (London: Vinton & Co., 1919), 116; Bell, "Breeds of Draft Horses," 10–1; Alvin Howard Sanders, *A History of the Percheron Horse* (Chicago, 1917), 181; *National Livestock Journal* 12 (Nov. 1881): 480; *Annual Report of the State Board of Agriculture of Ohio* (1890), 167; and *Annual Report of the State Board of Agriculture of Ohio* (1896), 623.

46. Robert Leslie Jones, *History of Agriculture in Ohio to 1850* (Kent, Ohio: Kent State University Press, 1983), 172; *Street Railway Journal* (Dec. 1885): 45; Carter, *Horses of the World,* 23; Alexander Easton, *A Practical Treatise on Street or Horse-Power Railways* (Philadelphia, 1859), 96; *Street Railway Journal* 4 (Oct. 1888), 268; C. B. Fairchild, "Trams: Electric, Cable, Horse and Other," *Street Railway Journal* 5 (Oct. 1890): 73; John Gilmer Speed, *The Horse in America: A Practical Treatise on the Various Types Common in the United States with Something of Their History and Varying Characteristics* (New York: McClure, Phillips & Co., 1905), 195; Gay, *Productive Horse Husbandry,* 310.

Economist Kyle Kauffman has attacked the traditional argument that owners preferred mules in the South because they handle heat better than horses do. He maintains that the choice was a function of ownership and agency. Mules, unlike horses, cannot be worked to death, so owners preferred them if they were being controlled by an agent (i.e., an employee) away from the owner's direct control. If employees sought to underfeed or beat mules, the animals resisted by refusing to work, so the owner's livestock was protected. See "Why Was the Mule Used in Southern Agriculture? Empirical Evidence of Principal-Agent Solutions," *Explorations in Economic History* 30 (1993): 336–51.

47. When we labeled the nineteenth-century urban horse as a technology on the electronic discussion group Envirotech, to our surprise we triggered a major flame war. See www.geocities.com/erech/animaltech.pdf (Aug. 2001).

48. William Cronon, *Nature's Metropolis: Chicago and the Great West* (New York: W. W. Norton, 1991), 48–54.

49. Lemuel Shattuck, *Report to the Committee of the City Council Appointed to Obtain the Census of Boston for the Year 1845* (Boston, 1846), 42; Commonwealth of Massachusetts, *Aggregate of Polls, Property and Taxes* (Boston, 1841). These figures are probably an undercount, since owners may have tried to hide horses from the tax collector. Also, the Boston metropolitan area was politically fragmented, so many horses that worked in the city lived in nearby towns.

50. U.S. Bureau of the Census, *Abstract of the Fourteenth Census of the United States, 1920* (Washington, D.C.: Government Printing Office, 1923), 123. The data were actually collected by county, although in each county an overwhelming number of humans lived in the city listed in this table. The figures for Kansas City may reflect the fact that it was a major transshipment center.

51. U.S. Census Office, *Twelfth Census of the United States, Taken in the Year 1900* (Washington, D.C., 1907), 428–79.

52. *NYT,* Mar. 20, 1879. Italics added.

One • Markets

1. Susan D. Jones, *Valuing Animals: Veterinarians and Their Patients in Modern America* (Baltimore: Johns Hopkins University Press, 2003), has the best analysis of commodification. The quotation is from Donald Worster, *Nature's Economy: A History of Ecological Ideas* (Cambridge: Cambridge University Press, 1985), 174. Worster contrasts this Victorian definition of civilization as the management of nature to the commonly held view that Victorians saw civilization as involving the survival of the fittest. He also points to Darwin's fondness for animals and Darwin's antivivisectionist views (181).

2. J. H. S. Johnstone, *Practical Treatise on the American Horse Breeding Industry as Allied to the Farm* (Chicago: Sanders Publ. Co., 1908), 108; Edward Mayhew, *Illustrated Horse Management* (Philadelphia, 1867), 357–65; Richard Mason, *The Gentleman's New Pocket Farrier* (Philadelphia: Gross & Elliot, 1826), 72; Isaac Lyon, *Recollections of an Old Cartman* (New York: New York Bound, 1984 reprint of 1872 ed.), 147; Henry C. Merwin, "The Horse Market," *Century* 49 (Mar. 1895): 694. The reminiscences of a livery stable keeper who recounts numerous tricks in buying, selling, and swapping horses are found in Upton Barnard, *Livery Stable Days* (San Antonio: Naylor Co., 1959), 233.

3. Keith Chivers, *The Shire Horse: A History of the Breed, the Society, and the Men* (London: J. A. Allen, 1976), 92.

4. John Stewart, *Stable Economy: A Treatise on the Management of Horses* (New York: D. Appleton, 1845), 327.

5. John B. Flinn, *Chicago: The Marvelous City of the West. A History, an Encyclopedia and a Guide* (Chicago: Flinn & Shepard, 1891), 197. Charles Wharton, *Handbook on the Treatment of the Horse in the Stable and on the Road* (Philadelphia: J. B. Lippincott, 1873), 17–21, has a series of illustrations that indicate the complexity of assessing age from the appearance of teeth.

6. Chivers, *The Shire Horse,* 101; Lyon, *Recollections of an Old Cartman,* 43; "Mrs. Vanderbilt's Hackneys," *Rider and Driver* (May 16, 1891): 2; *Annual Report of the State Board of Agriculture of Ohio* (1895), 135. J. H. S. Johnstone, *The Horse Book: A Practical Treatise on the American Horse Breeding Industry as Allied to the Farm* (Chicago: Sanders Publ. Co., 1908), notes a revival of interest in Cleveland Bays in 1908 and the importation of more stallions.

7. For examples of fraud in the horse market, see Lyon, *Recollections of an Old Cartman,* 47.

8. M. D. Hanover, *A Practical Treatise on the Law of Horses* (Cincinnati: Robert Clarke & Co., 1872), chs. 1 and 2, 57–99; Flinn, *Chicago,* 60; Johnstone, *The Horse Book,* 126; Herman Bidell, L. C. Douglas, Thomas Dykes, George Fleming, Archibald MacNeilage, Gilbert Murray, and W. R. Trotter, *Heavy Horses: Breeding and Management* (London: Vinton & Co., 1919), 154. Maurice Tellen, *The Draft Horse Primer* (Emmaus, Pa.: Rodale Press, 1977), 72–3, suggests that modern veterinary dentistry has made determining age more precise. *Rider and Driver* 2 (Aug. 19, 1893): 8, discusses Tattersall's. See also William Youatt, *The Horse with a Treatise on Draught and a Copious Index* (London: Baldwin & Craddock, 1861), 369. The varieties of fraud listed by Hanover are eye-opening examples of human cupidity. Youatt was published in seven different American editions, and this work is still held by twenty-eight different American libraries.

9. Flinn, *Chicago*, 19; Johnstone, *The Horse Book*, 118; Alfred M. Downes, *Firefighters and Their Pets* (New York: Harper & Brothers, 1907), 96; Edwin Emerson, "Making Policemen of Horses," *Harper's Weekly* 53 (Jan. 30, 1909): 27–8.

10. Youatt, *The Horse with a Treatise*, 369; Edward L. Anderson and Price Collier, *Riding and Driving* (New York: Macmillan Co., 1905), 31; "Editorial: The Blackmailing of Veterinarians by Horse Dealers," *American Veterinary Review* 22 (Nov. 1898): 517–9.

11. For such a large volume of horses, it made sense to bypass the middlemen (*New York Times* [hereafter cited as *NYT*], May 28, 1887). On regional variations in prices, see *Annual Report of the Chicago Board of Trade* (1893), 234.

12. John Harris, "Horses—Purchase, Disposition and Relative Value with Mules," *Street Railway Journal* 5 (Dec. 1887): 1043 (hereafter cited as *SRJ*); T. C. Barker, "Delayed Decline of the Horse in the Twentieth Century," in F. M. L. Thompson, ed., *Horses in European Economic History: A Preliminary Canter* (Reading: British Agricultural History Association, 1983), 105. The major decline in prices came after the electrification of street railways diminished demand. Horses disappeared faster from U.S. than British cities because of faster mechanization.

13. *Special Report on the Market for American Horses in Foreign Countries*, 55th Congress, 3rd session, Washington, 1898, 11; Flinn, *Chicago*, 279, 356.

14. Between 1892 and 1895, the number of horses in the United States declined from 16.2 million to 15.2 million.

15. *Annual Report of the Chicago Board of Trade* (1895), xxix; *Special Report on the Market for American Horses in Foreign Countries*, 55th Congress, 3rd Session, Washington, 1898, 9, 16, 19, 37, 49; Harold Sessions, *Two Years with the Remount Commissions* (London: Chapman & Hall, 1903), ch. 11; Flinn, *Chicago*, 72–3.

16. *The Stable: A Monthly Magazine Published for the Livery Stable, the Private Stable, and the Harness Saddlery* 1 (Feb. 1886): 49; "Stables and Care of Horses," *American Railroad Journal* 58 (1884): 306; Scott Molloy, *Trolley Wars: Streetcar Workers on the Line* (Washington, D.C.: Smithsonian Institution Press, 1996), 118.

17. Ervin F. Ewell, "The Fertilizing Value of Street Sweepings," *USDA Bulletin* 55 (1898): 17; Robert Orsi, *The Madonna of 115th Street: Faith and Community in Italian Harlem, 1880–1950* (New Haven: Yale University Press, 1985), 105.

18. Reginald S. Timmis, *Modern Horse Management* (Toronto and New York, 1921), 57; F. M. L. Thompson, "Horses and Hay in Britain," in Thompson, *Horses in European Economic History*, 64; Maurice Tellen, *The Draft Horse Primer* (Emmaus, Pa.: Rodale Press, 1977), 7; Marc Linder and Lawrence S. Zacharias, *Of Cabbages and Kings County: Agriculture and the Formation of Modern Brooklyn* (Iowa City: University of Iowa Press, 1999), 62; Ghislaine Bouchet, *Le Cheval à Paris de 1850 à 1914* (Geneva: Libraire Droz, 1993), 161. Bouchet found the volume of the freight on the farmers' morning (produce into the city) and afternoon (manure back to the farm) trips to be roughly equivalent.

19. C. B. Fairchild, "Trams: Electric, Cable, Horse and Other," *SRJ* 6 (Oct. 1890): 470; John H. White Jr., "Horse Power," *American Heritage of Science and Technology* 8 (Summer 1992): 61, 63. White estimated the manure value at six dollars per year for a thousand-pound horse, but urban draft animals could weigh twice that.

20. *NYT*, May 28 & Aug. 13, 1881; Linder and Zacharias, *Of Cabbages and Kings County*,

62; Albert H. Buck, *A Treatise on Hygiene and Public Health* (New York: William Wood, 1871), 466; Joel A. Tarr, *The Search for the Ultimate Sink: Urban Pollution in Historical Perspective* (Akron, Ohio: University of Akron Press, 1996), 293–9.

21. Albert H. Buck, *A Treatise on Hygiene and Public Health* (New York: William Wood, 1871), 466.

22. *NYT,* July 11, 1884; Jimmy Skaggs, *The Great Guano Rush: Entrepreneurs and American Overseas Expansion* (New York: St. Martin's Griffin, 1994), 141–8.

23. Jones, *Valuing Animals,* 6; American Society for the Prevention of Cruelty to Animals, *Annual Reports,* 1887–97. Its name notwithstanding, this was a New York City organization, hereafter referred to as ASPCA. Sigfried Giedeon, *Mechanization Takes Command: A Contribution to Anonymous History* (New York: Oxford University Press, 1948), 209–46, deals with the mechanization of death in the nineteenth century.

24. Zulma Steele, *Angel in Top Hat* (New York: Harper & Brothers, 1942), 68; *Annual Reports of the New York City Board of Health* are available at the website, "The Living City," http://156.145.78.54/htm/home.htm. The 1910 statistic is from the 1912 report, p. 12.

25. Lockwood and Everett, *Description of the Recent Improvements of Lockwood and Everett in Steam Rendering Apparatus* (New York, 1871), 19; U.S. Bureau of the Census, *Social Statistics of Cities,* 17 (1880), see unpaginated entry for New York City. The history of rendering in New York is probably best traced in the *Annual Reports of the Board of Health,* which are available in a word-searchable format at the website, "The Living City," http://156.145.78.54/htm/home.htm. For an interesting nuisance case involving a nineteenth-century rendering plant in Newark, N.J., see Donna J. Rilling, "Bone Boilers: Nineteenth Century Green Businessmen?" a paper presented at the 2005 annual meeting of the American Society of Environmental History in Houston. Jeff Kisselhoff, *You Must Remember This: An Oral History of Manhattan from the 1890s to World War II* (New York: Harcourt, Brace, Jovanovich, 1989), 29, 542, reports two interviews with men who recall playing with dead horses in their childhood. One reported jumping up and down on the carcass until the bladder burst after it had laid in the street for several days awaiting pickup by the humane society. But they were recalling the 1920s, when only a few thousand horses remained. Probably the old system for picking up horses had collapsed by then.

26. Bouchet, *Le Cheval à Paris,* 229–35.

27. *ASPCA Annual Reports* 2 (1868): 15; *New York Herald,* May 12, 1869; John S. Billings, *Relation of Animal Diseases to the Public Health and Their Prevention* (New York: D. Appleton & Co., 1884), 154. Dora was famous as the horse used in Edward Muybridge's well-known series of photos of horses in motion.

28. John Duffy, *A History of Public Health in New York, 1866–1966* (New York: Russell Sage, 1974), 2:377; Alan Downs, "Vicksburg," *Encyclopedia of the Confederacy* (New York: Simon & Schuster, 1993), 1660.

29. "The Dead Horse: How Its Carcass Is Utilized in the Manufacturing and Domestic Arts," *Journal of Comparative Medicine and Surgery* 17 (Aug. 1896): 597–8.

30. William P. Marchione, *The Bull in the Garden: A History of Allston Brighton* (Boston: Trustees of the Boston Public Library, 1986), 75; Samuel L. Dana, *A Muck Manual for Farmers* (Lowell, 1843), 27–41; Skaggs, *The Great Guano Rush,* 1; D. J. Browne, *The American Muck Book: Treating of the Sources, History, and Operations of All the Principal Fertilizers and*

Manures (New York: C. M. Saxton, 1851), 272; U.S. Census Bureau, *Social Statistics of Cities (1880)*, vol. 17. The census report has unpaginated entries for each city. Local case studies on these topics, especially for smaller and southern cities, are badly needed.

31. Lockwood and Everett, *Recent Improvements in Steam Rendering Apparatus*, 9; Browne, *The American Muck Book*, 107; Skaggs, *The Great Guano Rush*, 141–8; Barbara Rosenkrantz, *Public Health and the State: Changing Views in Massachusetts, 1842–1936* (Cambridge: Harvard University Press, 1972), ch. 9; Judith Walzer Leavitt, *The Healthiest City: Milwaukee and the Politics of Health Reform* (Princeton: Princeton University Press, 1982), 130–55.

32. *NYT*, Feb. 23, 1873; "John M. Tyler et al., petitioners, vs. John P. Squire et al., Respondents," *Official Record of the State Board of Health of Massachusetts: Together with a Phonographic Report of the Evidence and Arguments at the Hearing* (Cambridge, 1874), 4–212. This report contains 550 pages of eyewitness (or nosewitness?) complaints about "the big stink" and extensive expert testimony on the state of the art in pollution control. It is probably the best summary of the state of the art at that date. Lockwood and Everett, *Recent Improvements in Steam Rendering Apparatus*, describes one technology.

33. Department of the Interior, Census Office, *Eleventh Census, 1890: Population, Part 2*, 304–5. The census was maddeningly erratic in its definitions and the quality of its counting. In 1900 it did not count occupations at all. We should note that electrification was already well under way by this date. See *The Stable* 16 (Mar. 1887): 28–9, on the ratio of hostlers to horses.

34. Bureau of the Census, *Manufacturers 1905, Part IV, Special Reports on Selected Industries: Carriages and Wagons* (Washington, D.C.: Government Printing Office, 1908), 301–27.

35. Employment and firms in horse-related industries can be traced in the U.S. Census of Manufacturing. For the 1890 figures, see Bureau of the Census, *Report on Manufacturing Industries in the United States: 11th Census, Part III*, table 4: Manufacturers by Specified Industries: 1890 (Washington, D.C.: Government Printing Office, 1895), 92–3. For 1909, see U.S. Bureau of the Census, *Census of Manufactures* (Washington, D.C.: Government Printing Office, 1910), table 1: Comparative Summary for United States, by Industries, 1909, 1904, and 1899, 507–17. Most of these vehicles were not built for urban use, and it is impossible to separate out the urban portion. See Glenn Porter, ed., *Asher and Adams Pictorial Album of American Industry, 1876* (New York: Rutledge Books, 1976), 43, 46, for descriptions of whip companies.

36. U.S. Patent Office, *Patents for Inventions, 1790 to 1873 Inclusive* (Washington, D.C., 1874), 470–3; "How Horsecars Are Run," *The Stable* 2 (Mar. 1887): 28–9; *Boston City Directory*, 1900.

37. Francis B. Wilkie, *Davenport, Past and Present* (Davenport, Iowa: Luse, Lane & Co., 1858), 19–28; data from *Boston City Directory*, 1870–1900; *Pittsburgh Business Directory, 1892* (Pittsburgh, 1893).

38. *Moseman's Illustrated Guide for Purchasers of Horse Furnishing Goods* (New York: C. M. Moseman & Brother, 1892, reprinted London: Studio Editions, 1995).

39. *Livery: F. M. Atwood, Chicago* (New York: Rogers Peet & Co., 1904).

40. Nancy Kenoyer, "Wells Fargo Express Horses," *Western Horseman* (Apr. 1977): 58.

41. Figures based on crop production data in U.S. Department of the Census, *Census, 1900*, vol. 6, *Agriculture*, 21, 498. A higher estimate of the amount of land required to feed a horse can be found in T. C. Barker, "The Delayed Decline of the Horse in the Twentieth Century," in Thompson, *Horses in European Economic History*, 102–3.

42. U.S. Department of the Census, *Census, 1900*, vol. 5, *Agriculture*, 390.

43. *Sanborn Fire Insurance Map of Boston*, 1885.

44. Clay McShane, "La Construzione Sociale Della Strade in America" (translator unknown), in Mirko Zardini, ed., *Asfalto: Il Carattere della Citta* (Milan: Electa, 2003), 57–61.

Two • Regulation

1. J. B. DeBow, *Seventh Census of the United States, 1850* (Washington, D.C.: Robert Armstrong, Public Printer, 1853), cxvii–clxxix; Joseph C. G. Kennedy, *Census of the United States in 1860* (Washington, D.C.: Government Printing Office, 1864), 656–79; *Carriage Monthly* (Apr. 1904): 177; Annual Reports of Commonwealth of Massachusetts, *Aggregate of Polls, Property and Taxes*, 1870–1940.

2. For the increase of "control" in general in nineteenth-century technological cultures, see Miriam Levin, *Cultures of Control* (Amsterdam: Harwood Academic Publishers, 2000), and Phillip Thurtle, "Breeding and Training Bastards: Distinction Information and Inheritance in Gilded Age Trotting Horse Breeding," in Robert Mitchell and Philip Thurtle, eds., *Data Made Flesh: Embodying Information* (New York: Routledge, 2004), 65–86.

3. Graham Russell Hodges, *New York City Cartmen, 1667–1850* (New York: NYU Press, 1986), 2, 9, 18, 23, 46, 69, 143, 151–9, 171; John Jervis, *Horse and Carriage Oracle* (London: Henry Colburn, 1828), 174; Edward Mayhew, *Illustrated Horse Management* (Philadelphia: J. B Lippincott, 1867), 265. The assertion about the volume and types of vehicles is based on an examination of advertisements in the *Boston City Directory*, as well as those in I. N. Phelps Stokes, *The Iconography of Manhattan Island, 1498–1909* (New York: Arno Press, 1967, reprint of the Robert H. Dodd 1915 ed.).

4. Stokes, *The Iconography of Manhattan Island*, plates 66, 75; Hodges, *New York City Cartmen*, 2, 9, 18, 23, 69, 143, 151–9, 171; Thomas F. Devoe, *The Market Book* (New York: Burt Franklin, 1862); *New York Times* (hereafter cited as *NYT*), Mar. 20 & Apr. 28, 1878; Timothy F. Kruse, "Teamsters in the Gilded Age," M.A. thesis, University of Wisconsin–Madison, 1970, 11. Carl Bridenbaugh, *Cities in Revolt: Urban Life in America, 1743–1776* (New York: Knopf, 1955), 244, has a description of the process. The idea of licensed carts persisted as a restrictive measure. As late as 1878, New York City haulers sought licensing that would ban Brooklyn residents from bringing their carts into the city.

5. Lyon once owned a horse that had become nearly immobile, even when whipped, so he took him to a friend who was a jockey/horse doctor/cartman. When folk medicine didn't work, he took the twenty-year-old horse to an auctioneer, who dyed him, dosed him with pepper to make him livelier, and sold him as a nine-year-old for $30. The horse died three days later. See Isaac Lyon, *Recollections of an Old Cartman* (New York: New York Bound, 1984 reprint of the 1872 ed.), 13. On Pittsburgh, see H. Thurston, *Allegheny City's One Hundred Years* (Pittsburgh: A. A. Anderson & Son, 1888).

6. Lyon, *Recollections of an Old Cartman*, viii, 3–8, 43–49; Clay McShane, *Down the Asphalt Path* (New York: Columbia University Press, 1994), 44.

7. Francis M. Ware, *Driving* (New York: Doubleday, Page & Co., 1903), 106; "Mules," in Charles Reason Wilson and William Ferris, eds., *Encyclopedia of Southern Culture* (Chapel Hill: University of North Carolina Press, 1989), 511; Paul Underwood Kellog, ed., *Wage Earning Pittsburgh* (New York: Survey Associates, 1914), 121; Scott Molloy, *Trolley Wars: Streetcar Workers on the Line* (Washington, D.C.: Smithsonian Institution Press, 1996), 161; Edwin Valentine Mitchell, *The Horse and Buggy Age in New England* (New York: Howard McCann, 1937), 123; Roland L. Freeman, *The Arabbers of Baltimore* (Centreville, Md.: Tidewater Publishers, 1989), 5.

8. "Relations between Labor and Capital," Hearings of the Senate Committee on Education and Labor (1883), 772; John R. Commons, "Types of American Labor Organizations: The Teamsters of Chicago," *Quarterly Journal of Economics* 19 (1905): 400–33; David Witwer, "Unionized Teamsters and the Struggle over the Streets of the Early Twentieth Century City," *Social Science History* 24 (Spring 2000): 1–3.

9. Edward L. Anderson and Price Collier, *Riding and Driving* (New York: Macmillan Co., 1905), 365; Belle Beach, *Riding and Driving for Women* (New York, 1912), 267; Ware, *Driving*, 56, 62–8, 98–104; Anderson and Collier, *Riding and Driving*, 316; Wells Fargo & Co. Express, *Rules and Instructions* (San Francisco, 1902), n.p.; W. E. Partridge, "Rudiments of Driving," *Street Railway Journal* 3 (Nov. 1886): 16; Augustine Wright, *American Street Railways: Their Construction, Equipment and Maintenance* (Chicago, 1888), 122.

10. See the images at http://memory.loc.gov/ammem/index.html: A Trip down Market St. before the Fire (San Francisco, 1905), Lower Broadway (New York City, 1901), Union Square (New York City, 1901), Herald Square (New York City, 1906), A Visit to Berkeley (1906), Parade of Horses on Speedway (New York City, 1906), South Spring St. (Los Angeles, 1898), all in the American Memory Collection of the Library of Congress website, visited June 29, 2002.

11. Anderson and Collier, *Riding and Driving*, 318.

12. *The Charter and Ordinances of the City of Boston* (1856), 101.

13. "Forty Horses Unwinding a Cable Rope," *Street Railway Journal* 3 (Dec. 1887): 1040.

14. The New York City Fire Department claimed that it could harness a horse in 8.5 seconds. See *Social Statistics of Cities, 1880 Census*, vol. 17. This census consists of unpaginated reports from different cities. Alfred M. Downes noted that New York firemen, like other drivers, were closely bonded to their animals: "Great is the fire horse and mighty is his master's love and respect for him. Any needless injury or neglect, any annoyance or teasing, or even a suggestion of brutality, to the noble animal is regarded as a crime worthy of the severest punishment. The horses are the firemen's pride, their chums in leisure hours and their partners in danger." Alfred M. Downes, *Firefighters and Their Friends* (New York: Harper & Brothers, 1907), 11, 95; Edward Edwards, *History of the St. Louis Fire Department* (St. Louis: Central Publishing Co., 1916), 199–206.

15. Edwin Emerson, "Making Policemen of Horses," *Harper's Weekly* 53 (Jan. 30, 1909): 27–8; Stefan Lorant, *Pittsburgh: The Story of an American City* (Lenox, Mass.: Author's Editions, 1964), 177.

16. Molloy, *Trolley Wars*, 119; Alexander Easton, *A Practical Treatise on Street or Horse-*

Power Railways (Philadelphia, 1859), 105; Wright, *American Street Railways,* 116, 177; *City Work Horses,* Bulletin 173 (Storrs, Conn.: Agricultural Experiment Station, May 1931), 10. For a detailed description of training horses to back up, see Anderson and Collier, *Riding and Driving,* 135–7.

17. Wells Fargo & Co. Express, *Rules and Instructions for Drivers* (San Francisco, 1902, 1914), n.p. Wells Fargo also required drivers of "money wagons," the ancestors of today's armored cars, to carry pistols.

18. Witwer, "Unionized Teamsters," 10–1.

19. Mildred Walker, *The Brewer's Big Horses* (New York: Harcourt, Brace, 1940), 6.

20. Theodore Dreiser, *The Titan* (New York: World Publishing Co., 1951), 19.

21. Anderson and Collier, *Riding and Driving,* 98–9.

22. James Garland, *The Private Stable: Its Establishment, Management, and Appointments* (Boston: Little Brown, 1899), 578.

23. Hodges, *New York City Cartmen,* 16, 67, 97; Kruse, "Teamsters in the Gilded Age," 82.

24. "Relations between Labor and Capital," *Hearings of the Senate Committee on Education and Labor* (Washington, D.C., 1883), 771.

25. *The Charter and Ordinances of the City of Boston* (1856), 89; Peter Baldwin, *Domesticating the Street: Public Space in Hartford, 1850–1920* (Columbus: Ohio State University Press, 1999), 182; Barry Temkin, "Oats and Iron: Horse Drawn Transportation in Milwaukee, 1850–1890," B.A. thesis, University of Wisconsin, 1970, 9.

26. As quoted in Kruse, "Teamsters in the Gilded Age," 2.

27. Albert H. Buck, *A Treatise on Hygiene and Public Health* (New York: William Wood, 1871), 55.

28. Commons, "The Teamsters of Chicago," 401–33. Timothy Kruse's excellent 1990 University of Wisconsin M.A. thesis, "Teamsters in the Gilded Age," 4, argues against Commons's comments about teamsters' unions being like guilds, suggesting that the overwhelming majority of teamsters were employees by the end of the nineteenth century. Kruse is correct in terms of sheer numbers, but the smaller number of owners could and did produce guildlike unions.

29. *R. G. Dun Credit Reports,* Baker Business Library, Harvard University. The fate of five firms is unknown.

30. Commons, "The Teamsters of Chicago," 423; Kruse, "Teamsters in the Gilded Age," 26. Kruse also points out an occasion when coal drivers, who had heavy wagons and were underemployed in the summer, if employed at all, broke a New York teamsters strike involving construction work—the short construction work season gave strikers great leverage at this time of the year.

31. T. J. Minihan, "New York City," *International Teamster* 2 (May 1905): 19.

32. Molloy, *Trolley Wars,* 18. Rush hours created widely separated periods of demand with quite variable work hours. James McCabe, *New York by Sunlight and Gaslight* (Philadelphia: Hubbard Brothers, 1882), 143, notes eighteen-hour split shifts on New York transit lines.

33. Kruse, "Teamsters in the Gilded Age," 58; International Brotherhood of Teamsters, *Teamsters All: Pictorial Highlights in Our History* (Washington, D.C., 1976), 34. The teamsters did make some efforts to organize blacks but mostly in southern cities.

34. For that reason the job of driving for the railway express was highly valued—see the description of one fictional worker in James T. Farrell, *A World I Never Made* (New York: Vanguard Press, 1936), 380–2.

35. Commons, "The Teamsters of Chicago," 408–15.

36. "The Horse," *Nation* 60 (June 20, 1895): 476.

37. McCabe, *New York by Sunlight and Gaslight,* 143.

38. Farrell, *A World I Never Made,* 380–2; James T. Farrell, "A Teamster's Payday," in *More Stories* (Garden City, N.Y.: Sundial Press, 1944), 58.

39. Henry Roth, *Call It Sleep* (Paterson, N.J.: Pageant Books, 1960 reprint of the original 1933 ed.), 247.

40. On benefit associations, see McCabe, *New York by Sunlight and Gaslight,* 143.

41. Freeman, *The Arabbers of Baltimore,* 25–67, quotation on 33.

42. Roth, *Call It Sleep,* 378, also 369. Carol Zisowitz Stearns and Peter N. Stearns, *Anger: The Struggle for Emotional Control in America's History* (Chicago: University of Chicago Press, 1986), 31. The authors speculate, probably accurately, that capitalism may have required the "smiling shopkeeper's personality," since the factory system can't operate amid intense worker rage. The same Enlightenment ideals that tolerated complete exploitation of animals assumed that humans were rational and hence capable of controlling rage. Katherine C. Grier, "Animal House: Pet-keeping in Urban and Suburban Households in the Northeast, 1850–1900," in Peter Benes, ed., *New England's Creatures, 1400–1900* (Boston: Boston University Press, 1996), 117, notes that Victorians encouraged beekeeping by children to discourage their innate cruelty. See also Roy Rosenzweig and Betsy Blackmar, *The Park and the People: A History of Central Park* (Ithaca: Cornell University Press, 1992), 223.

43. Bergh's ideas often seem inconsistent. For example, he favored hippophagy, since he believed that the consumption of horsemeat would take older, lamer animals off the street faster. See *ASPCA Annual Report* 2 (1868): 15, and *New York Herald,* May 12, 1869. The opponent of the lash for horses also favored whipping some human criminals. Apparently, he was also that great American rarity, a monarchist (see file 1 in the ASPCA Archives). The Executive Committee, according to the 1868 *Annual Report,* included some of the wealthiest individuals in Manhattan high society, many of them with horse-dependent businesses or involved in leisure use of the horse. Among the committee members were magazine publisher Frank Leslie and newspaper publisher James Gordon Bennett, who provided plentiful publicity for the organization. See, e.g., *Frank Leslie's Magazine,* Oct. 28 & Dec. 9, 1865, and May 5, 1866, and numerous *New York Herald* articles in the Clippings File in the ASPCA Archive.

44. See the June 24, 1871, letter from Mayor A. Oakley Hall to the New York City Police Commission (copy in the ASPCA Archives), demanding an end to Bergh's arrests and removal of horses. The mayor wanted a warrant issued and judicial process, not "summary judgment on the streets." Bergh responded with a complaint about "rich and powerful corporations." Hall, a Tammany mayor, may have been speaking for the teamsters, who were more often the focus of Bergh's activities. Hall did acknowledge the general justice of Bergh's cause.

45. Vanderbilt was initially receptive to Bergh, sending one driver over to apologize,

firing another, and agreeing in principle to fire cruel drivers, but by 1868 Vanderbilt was no longer replying to Bergh's letters. ASPCA Archives, letters of May 3 & Oct. 2, 1866, and undated letters, folder 3 (1868).

46. *Journal of Commerce,* July 7, 1867. Salting streets was a contested issue with heavy overtones of social class. Recreational horse users, including those on the ASPCA board, believed that salting snow damaged their horse's fetlocks. Moreover, melting snow limited sleighing, a favorite winter pastime for the carriage set. Streetcar companies often debated the use of salt—freight companies and teamsters wanted streets salted, since wagons couldn't move until the snow was gone. In a few cases they kept a stock of freight sleds or even omnibus sleighs for the winter. See "Science in Familiar Things: Salting the Streets to Remove Snow—Its Effect upon Horses' Hoofs," *Scientific American* 2 (Jan. 14, 1860): 41.

47. *ASPCA Annual Report* 3 (1867): n.p. The arrest of the president of the Bleecker Street Railway is described in a clipping from an unidentified newspaper for February 15, 1871, in the ASPCA clipping file for that year. Both August Belmont and A. T. Stewart announced their opposition at board meetings. See the minutes of the May 12, 1866, meeting in the ASPCA archives. The quotation about the "fine span of horses" is from the *Journal of Commerce,* July 27, 1867.

48. Allegheny County Humane Society (hereafter cited as ACHS), *Second Annual Report* (1874), 12–3; *Fifth Annual Report* (1880), 7, 28. Humane groups in other cities seem to have followed this general pattern; Western Pennsylvania Humane Society, *Forty-Second Annual Report* (1915). This report contains a history of the organization, with a list of its early directors.

49. Bergh to Dodd's Express, Sept. 10, 1866, ASPCA Files, New York City.

50. See the January 1869 photo showing a fire being set under a horse while he is being hit with a stick, in file 3, ASPCA Archives. The same folder contains other grotesque illustrations.

51. *Eleventh Annual Report of the Board of Mediation and Arbitration* (New York, 1898), as quoted in Kruse, "Teamsters in the Gilded Age," 81.

52. *NYT,* Jan. 4, 1873: "Bergh's Agent has stopped a 2600 pound one horse car, ordering men and boys out. He stopped another two horse car with 137 passengers." See also *New York World,* Jan. 5, 1887.

53. Wells Fargo & Co. Express, *Rules and Instructions* (San Francisco, 1902). It is hard to believe that this rule was enforced.

54. The quotation is from ACHS, *Forty Third Annual Report* (1919), 20.

55. Boston Work Horse Parade Association, *Seventh Annual Report* (1909), 5. The association was funded by the Massachusetts SPCA.

56. ASPCA *Annual Report* 8 (1874), 18.

57. Paul Pinkerton Foster, "Helping the Work Horses," *Outing* 53 (May 1919): 168–79.

58. Kruse, "Teamsters in the Gilded Age," 70; Molloy, *Trolley Wars,* 120–1; "Shying Horses," *International Teamster* 3 (Jan. 1906): 28; "Horse Sense Reminder," *International Teamster* 3 (Jan. 1906): 15. "Horse Sense Reminder" was a Black Beauty–like piece, written in the first person by an imaginary horse who reminded teamsters, among other things, that he did not like to gallop downhill. *Shying* is when a horse suddenly jumps one way or the other and is usually caused by the horse being frightened.

59. Molloy, *Trolley Wars,* 116.

60. ACHS, *First Annual Report* (1874), 30.

61. Freeman, *The Arabbers of Baltimore,* 9.

62. "Report of the Committee on Buildings," *Verbatim Report of the Proceedings of the Convention Relative to the Organization of the American Street Railway Association* (Boston, 1882), 51.

63. Boston Work Horse Parade Association, *Seventh Annual Report* (1909), 113 et seq.; "Various Vehicles," *The Stable* 2 (July 1987): 112; interview with Anthony Penna, Nov. 1, 2003; "How Horsecars Are Run," *The Stable* 2 (Mar. 1887): 28–9; ACHS, *Twenty-fifth Annual Report* (1924), 5–6; Paul Pinkerton Foster, "Helping the Work Horses," *Outing* 53 (May 1919): 68–79; Reginald S. Timmis, *Modern Horse Management* (Toronto and New York, 1921), 12.

64. Garland, *The Private Stable,* 578; Timmis, *Modern Horse Management,* 44; John Gilmer Speed, *The Horse in America: A Practical Treatise on the Various Types Common in the United States with Something of Their History and Varying Characteristics* (New York: McClure, Phillips & Co., 1905), 46; New York City Board of Health, *Annual Reports,* as cited in McShane, *Down the Asphalt Path,* 175.

65. On the origins of keeping to the right (and for that matter keeping to the left in societies where that was the norm), see Maxwell Lay, *Ways of the World: A History of the World's Roads and the Vehicles That Use Them* (New Brunswick, N.J.: Rutgers University Press, 1992), 197–201. See the American Memory Collection of the Library of Congress website for images of street traffic. Francis V. Greene, "An Account of Some Observations of Street Traffic," *American Society of Civil Engineers Transactions* 15 (1886): 123–38, has the traffic counts. Another source claimed 20,000 vehicles a day for Broadway in 1882. See McCabe, *New York by Sunlight and Gaslight,* 143. Greene, an engineer, seems more credible than McCabe, a sensationalistic journalist.

66. City of Detroit, *Ordinances and By-Laws* (1855); Joseph E. Gary (comp.), *Laws and Ordinances of the City of Chicago, January 1, 1866* (Chicago, 1866); *NYT,* July 26, 1865. In *Social Statistics of Cities, 1880 Census,* vols. 17 and 18, reporters from several cities (entries are unpaginated) summarize their regulations. See McShane, *Down the Asphalt Path,* for a summary of the 1890 Census data.

Three • Powering Urban Transit

1. Henry C. Binford, *The First Suburbs: Residential Communities on the Boston Periphery, 1815–1860* (Chicago: University of Chicago Press, 1985), 84–8.

2. The philosopher Blaise Pascal had operated a short-lived predecessor from 1662 to 1677.

3. Nicholas Papayanis, *Horse-Drawn Cabs and Omnibuses in Paris: The Idea of Circulation and the Business of Public Transit* (Baton Rouge: Louisiana State University Press, 1996), 16–86; Roy Shadwell, *Horse Omnibus* (London: Author, 1994), 4–5.

4. T. C. Barker and Michael Robbins, *A History of London Transport: Passenger Travel and the Development of the Metropolis,* 2 vols. (London: George Allen & Unwin, 1963), 1:1–24.

See also W. J. Gordon, "The Horse World of London: The Omnibus Horse," *Leisure Hour* 41 (Nov. 1891): 27–32.

5. George Rogers Taylor, "The Beginnings of Mass Transportation in Urban America: Part I," *Smithsonian Journal of History* 1 (1966): 41.

6. John H. White Jr., *Horsecars, Cable Cars and Omnibuses: All 107 Photos from the John Stephenson Company Album, 1888* (New York: Dover Publications, 1974), vii–viii. A picture of the New York car designed by John Stephenson can be found on p. vii. See also William Louis Gannon, "Carriage, Coach and Wagon: The Design and Decoration of American Horse-Drawn Vehicles," Ph.D. diss., Iowa State University, 1960, 108–9; Ezra M. Stratton, *The World on Wheels* (New York: B. Blom, 1878), 432–9; Taylor, "The Beginnings of Mass Transportation: Part I," 39–42; and Clay McShane, "Transforming the Use of Urban Space: A Look at the Revolution in Street Pavements, 1880–1924," *Journal of Urban History* 5 (May 1979): 279–307.

7. Taylor, "The Beginnings of Mass Transportation: Part I," 44–5.

8. Frederic W. Speirs, *The Street Railway System of Philadelphia: Its History and Present Conditions*, Johns Hopkins University Studies in Historical and Political Science, 15th Series (Baltimore: Johns Hopkins University Press, 1897), 10–11.

9. Taylor, "The Beginnings of Mass Transportation: Part I," 45–6; Binford, *The First Suburbs*, 89–90. Aside from New Orleans and Washington, D.C., no city under 40,000 population had a line before 1840. Baltimore was the other extreme, with 102,000 people in 1840 but no omnibus until 1844. See Arthur J. Krim, "The Development of Internal Urban Transport in North America, 1830–1860," paper delivered at the American Association of Geographers Convention, Kansas City, Apr. 1972, 7–8.

10. Krim, "Development of Internal Urban Transport," 7–10, tables A, B, C.

11. Ibid., 5.

12. Dr. Waterbury, "Feeding of State Horses in New York City," *Transactions of the American Institute* (1855): 466–9.

13. W. Gilmore Sims, "The Philosophy of the Omnibus," *Godey's Ladies Book* 13 (Sept. 1841): 104.

14. "Notice to Omnibus Drivers and Others," *Brooklyn Daily Eagle* [hereafter cited as *BE*], Aug. 10, 1850. Those who informed authorities about speeding violations were entitled to one-half of the penalty.

15. The Brooklyn "omnibus war" is discussed in *BE*, July 17 & 30, 1845; Nov. 25, 28 & Dec. 1, 15, 17, 1846. For omnibus operations, see the following two articles in the *New York Times* [hereafter cited as *NYT*]: "The Departing Omnibus: Its Prevalence in New York Thirty Years Ago," Nov. 13, 1881, and "More about the Omnibus," Dec. 4, 1881.

16. Alexander Easton, *A Practical Treatise on Street or Horse-Power Railways: Their Location, Construction, and Management* (Philadelphia: Crissy & Markley, 1859), 5.

17. James D. McCabe, *New York by Sunlight and Gaslight* (Philadelphia: Hubbard Brothers, 1882), 143, 238. Photographs of omnibuses, some quite elaborate, can be found in White, *Horsecars, Cable Cars and Omnibuses*, plates 11–17.

18. Nicholas B. Wainwright, ed., *A Philadelphia Perspective: The Diary of Sidney George Fisher, 1834–1871* (Philadelphia: Historical Society of Pennsylvania, 1967), 316; Frank Row-

some Jr., *Trolley Car Treasury: A Century of American Streetcars—Horsecars, Cable Cars, Interurbans, and Trolleys* (New York: Bonanza Books, 1956), 19. For complaints about omnibus manners and overloading, see the following articles in *BE:* "Manners in New York," Aug. 20, 1846; "The Man Who Smokes in the Omnibus," May 11, 1850; "Omnibus Etiquette," Aug. 7, 1850; "Overloading Omnibusses [*sic*]," June 24, 1853. See also "Omnibuses as a School of Manners," *Leisure Hour* 35 (1886): 135.

19. Quoted in Edward K. Spann, *The New Metropolis: New York City, 1840–1857* (New York: Columbia University Press, 1981), 285–6.

20. McCabe, *New York by Sunlight and Gaslight,* 143.

21. Joel A. Tarr, *Transportation Innovation and Changing Spatial Patterns in Pittsburgh, 1850–1934* (Chicago: Public Works Historical Society, 1978), 5. These disputes over omnibus stands also took place in other cities. See, e.g., "Omnibus War," *BE,* Dec. 17, 1846.

22. Papayanis, *Horse-Drawn Cabs and Omnibuses,* 62–3; W. J. Gordon, "The Horse World of London: The Omnibus Horse," *Leisure Hour* 41:27.

23. Quoted in Taylor, "The Beginnings of Mass Transportation: Part I," 47–8.

24. Sims, "Philosophy of the Omnibus," 106. One British writer took a different position, maintaining that the mixing of classes on the omnibus had a "refining influence" because of the example set by the "well-bred." See also "Omnibuses as a School of Manners," 135–7.

25. Binford, *The First Suburbs,* 90; Taylor, "The Beginnings of Mass Transportation: Part I," 48.

26. Quoted in Taylor, "The Beginnings of Mass Transportation: Part I," 48. See also Spann, *The New Metropolis,* 285.

27. *BE,* Sept. 8, 1849.

28. Joseph Warren Greene Jr., "New York City's First Railroad, the New York & Harlem, 1832 to 1867," *New York Historical Society Quarterly Bulletin* 9 (Jan. 1926): 107–17.

29. White, *Horsecars, Cable Cars and Omnibuses,* 21. Stephenson named the car the John Mason, in honor of the railroad's president.

30. Quoted in Greene, "New York City's First Railroad," 110. See also Edwin G. Burrows and Mike Wallace, *Gotham: A History of New York City to 1898* (New York: Oxford University Press, 1999), 565.

31. Spann, *The New Metropolis,* 289, 292; Taylor, "The Beginnings of Mass Transportation in Urban America: Part II," 1 (Autumn 1966): 34–5.

32. Harry James Carman, *The Street Surface Railway Franchises of New York City* (New York: Columbia University Press, 1919), 19–26; Greene, "New York City's First Railroad," 119; *NYT,* May 17, 1852.

33. Spann, *The New Metropolis,* 289, 292; Taylor, "The Beginnings of Mass Transportation: Part II," 34–5.

34. Spann, *The New Metropolis,* 289; Robert C. Brooks, "History of the Street and Rapid Transit Railways of New York City," Ph.D. diss., Cornell University, 1903; John Noble & Co. (comp.), *Facts Respecting Street Railways; The Substance of a Series of Official Reports from the Cities of New York, Brooklyn, Boston, Philadelphia, Baltimore, Providence, Newark, Chicago, Quebec, Montreal, and Toronto* (London: P. S. King, 1866), 15, 21; "The Way to Wealth for

Our City Is on the Railways within the City," *American Railroad Journal* 26 (May 7, 1853): 296–7.

35. In 1885, the Broadway omnibus companies sold off their omnibuses to be used in smaller towns and cities. The *New York Times* observed that "once the omnibus owned our streets. . . . Now, like the Indian, it is fast disappearing, and has become an object of curiosity and contempt to staring countrymen." See "Lo! The Poor Omnibus," *NYT*, July 12, 1885; Noble, *Facts Respecting Street Railways*, 15; and "How We Ride in New York," *American Railroad Journal* 36 (Nov. 11, 1863): 1047.

36. Compiled from Office of the Census, *Tenth Census: Social Statistics of Cities* (1880).

37. New York State Engineer, *Annual Report*, 1855, 186, 224, 659; White, *Horsecars, Cable Cars and Omnibuses*, xi; "How We Ride in New York," 1047; *Street Railway Journal* 3 (May 1886): 503; John H. White Jr., "Horse Power," *American Heritage of Science and Technology* 8 (Summer 1992): 41–51; Donald L. Berkebile, *Carriage Terminology: An Historical Dictionary* (Washington, D.C.: Smithsonian Institution, 1978), 74. Information on vehicle weight was too sporadic to allow consistent time series data.

38. White, "Horse Power," 49; White, *Horsecars, Cable Cars and Omnibuses*, vii–xi.

39. "Letter from John Stephenson & Co.," *American Railroad Journal* 57 (June 15, 1883): 36; White, *Horsecars, Cable Cars and Omnibuses*, xi; White, "Horse Power," 50.

40. Easton, *Practical Treatise on Horse-Power Railways*, 5.

41. Tarr, *Changing Spatial Patterns in Pittsburgh*, 7.

42. Noble, *Facts Respecting Street Railways*, 16.

43. "The Growth of City Traffic," *Engineering News*, Oct. 15, 1887, 273–5.

44. Data compiled from the Office of the Census, *Tenth Census: Social Statistics of Cities* (1880), and *Eleventh Census, Statistics of Transportation*. Other cities with large increases were St. Louis, from 56 to 150 rides per capita, an increase of 168%, and San Francisco, 101 to 270 rides per capita, an increase of 167%.

45. Spann, *The New Metropolis*, 295.

46. Binford, *The First Suburbs*, 146–8; Walter S. Allen, *Street Railways: Development of Street Railways in the Commonwealth of Massachusetts* (Boston: Commonwealth of Massachusetts, n.d.), 3–5.

47. Noble, *Facts Respecting Street Railways*, 21. In some cases omnibus owners obtained street railway franchises—it was relatively cheap for them to go into the business, since they already had horses and stables.

48. Sam Bass Warner Jr., *Streetcar Suburbs: The Process of Growth in Boston, 1870–1900* (Cambridge: Harvard University Press, 1962); "Boston," *NYT*, Aug. 31, 1853. Before 1873, some professionals, such as lawyers, had begun commuting from outside the city. Warner doesn't examine the major landfills in the Back Bay and South End, which not only created new land for housing but opened routes to outlying areas to the south that once had been accessible only by a narrow neck. Binford, *The First Suburbs*, 147–9, suggests a greater effect on Cambridge and Somerville, just west of Boston. Boston could not expand much to the north or east because of its harbor.

49. Massachusetts State Railroad Commission, *Annual Reports*.

50. Warner, *Streetcar Suburbs*, 25–6.

51. The material on Pittsburgh is taken from Tarr, *Changing Spatial Patterns in Pittsburgh,* 1–14.

52. Annexation of contiguous townships and boroughs added approximately seventy-five thousand people.

53. Tarr, *Changing Spatial Patterns in Pittsburgh,* 13. For white-collar employment and commuting, see Ileen A. DeVault, *Sons and Daughters of Labor: Class and Clerical Work in Turn-of-the-Century Pittsburgh* (Ithaca: Cornell University Press, 1990), 146. DeVault is primarily discussing the use of the electric streetcar, but there is considerable evidence that such commuting patterns began in the horsecar era.

54. John Henry Hepp IV, *The Middle-Class City: Transforming Space and Time in Philadelphia, 1876–1926* (Philadelphia: University of Pennsylvania Press, 2003), 9, 26–7.

55. Office of the Census, *Social Statistics of Cities: 1880* has an excellent collection of maps.

56. Wainwright, *A Philadelphia Perspective,* 316.

57. Taylor, "The Beginnings of Mass Transportation: Part II," 47–9; Joel Schwartz, "'To Every Mans Door': Railroads and Use of the Streets in Jacksonian Philadelphia," *Pennsylvania Magazine of History and Biography* 128, no. 1 (Jan. 2004): 39. In Philadelphia, e.g., the City Railroad, created in December 1834 and initially running for a seven-block area, primarily delivered freight.

58. "Report on London Transport," *Street Railway Journal* 2 (July 1886): 480; "Report on Paris Transport," *Street Railway Journal* 2 (July 1886): 483. Omnibuses were a more viable mode in both cities, since their streets were better paved than American cities.

59. Noble, *Facts Respecting Street Railways.* There was also opposition in St. Louis based on fear of "monopolization of the public street by a private company for private gain." See Andrew C. Young, "The St. Louis Streetcar Story," *Interurbans Special 108* (Glendale, Calif.: Interurban Press, 1972).

60. Objections by Philadelphians to street railways are enumerated in Speirs, *Street Railway System of Philadelphia,* 12–13, and Binford, *The First Suburbs,* 89–90. The *Brooklyn Daily Eagle* also reported popular opposition to streetcars on certain streets. See *BE,* Feb. 2 & 16, 1864.

61. Michael Bruce Kahan, "Pedestrian Matters: The Contested Meanings and Uses of Philadelphia's Streets, 1850s-1920s," Ph.D. diss., University of Pennsylvania, 2002, 36–41. We thank Dr. Kahan for providing a copy of his dissertation.

62. Schwartz, "To Every Mans Door," 35.

63. Quoted in Kahan, "Pedestrian Matters," 39–42.

64. Ibid., 43–5.

65. Quoted in Speirs, *Street Railway System of Philadelphia,* 17.

66. Kahan, "Pedestrian Matters," 47, 52.

67. Ibid., 52.

68. Joel A. Tarr and Josef Konvitz, "Patterns in the Development of the Urban Infrastructure," in Howard Gillette and Zane Miller, eds., *American Urbanism: A Historiographical Review* (Westport, Conn.: Greenwood Press, 1987), 195–226.

69. Wainwright, *A Philadelphia Perspective,* 316. Fisher made a similar comment in his entry for July 14, 1859. See pp. 327–8.

70. In most municipal streetcar charters, the firms were held responsible for paving and repairing streets between and within a certain distance of their tracks. Companies and the municipalities frequently argued over proper street maintenance. The Philadelphia experience with this requirement is discussed in Speirs, *Street Railway System of Philadelphia*, 54–64. See also Robin L. Einhorn, *Property Rules: Political Economy in Chicago, 1833–1872* (Chicago: University of Chicago Press, 1991), 217–24. Einhorn notes that the "segmented approach" to city building, where abutters had power to determine whether they wished an improvement, was rejected from the beginning in Chicago. Partial compensation initially came through ordinances that required streetcar companies to pay for one-third of street paving costs, but even this was voided through a 1869 court decision.

71. For discussions of problems with various types of rails, see "Width of Carriage Tracks," *Street Railway Journal* 2 (Jan. 1886): 78; Tom L. Johnson, *Street Railroad Construction* (Louisville, Ky.: Chas. E. Dearing, 1883); James R. Alexander, "Technological Innovation in Early Street Railways: The Johnson Rail in Retrospective," *Railroad History* 164 (1991): 64–85. We thank Professor Alexander for providing us with copies of his publications on rails. T-rails used for steam railroads were not suited for city streets and were banned in some cities. See Michael Massouh, "Technological and Managerial Innovation: The Johnson Company, 1883–1898," *Business History Review* 50 (Spring 1976): 49n8.

72. "Collision on the Fulton Street Railtrack," *BE*, July 19, 1855.

73. Wright, *American Street Railways*, 11–12. As late as 1904, a study by engineer John Alvord found that most private vehicles in Chicago traveled on the trolley tracks. See Paul Barrett, *The Automobile and Urban Transit: The Formation of Public Policy in Chicago, 1900–1930* (Philadelphia: Temple University Press, 1983), 52.

74. "Our City Railroads," *NYT*, Apr. 25, 1859. See also Q. A. Gillmore, *A Practical Treatise on Roads, Streets, and Pavements*, 9th ed. (New York: D. Van Nostrand, 1896), 234; Alexander, "Technological Innovation in Early Street Railways," 71.

75. Tubal Cain, "City Railroads: What City Railroads Might Be—Necessity for an Improved Pavement," *NYT*, Jan. 6, 1860, and "Street Railways: How They Should Be Constructed," *NYT*, Mar. 26, 1860. Tubal Cain was actually Alexander Holley, who at this time was deeply involved in railroad engineering. He had worked at the New York Locomotive Works, was part owner and editor of *Railroad Advocate*, a short-lived railroad journal, and had studied and reported on European railway practice and engineering. He supposedly had written at least 276 articles for the *New Yrok Times*, on a range of engineering and other topics. Holley later became famous as the nation's foremost builder of Bessemer furnaces. For Holley's career, see Jeanne McHugh, *Alexander Holley and the Makers of Steel* (Baltimore: Johns Hopkins University Press, 1980). Holley as "Tubal Cain" is discussed in note 30, p. 59.

76. Strickland Kneass to J. M. Gibson, Esq., Oct. 12, 1855, reprinted in Easton, *Practical Treatise on Horse-Power Railways*, 55–60.

77. Noble, *Facts Respecting Street Railways*, 10–12.

78. Johnson had previously invented a fare box that became the industry standard. Massouh, "Technological and Managerial Innovation," 47–54; James R. Alexander, "Technological Innovation in Steel Fabrication: Marketing and Production Considerations in the Manufacture of the Johnson Rail," *Business and Economic History* 20 (1991): 210–5.

79. Wright, *American Street Railways,* 45–54. The streetcar company in rainy Shreveport, La., laid its crossties and rails on unpaved streets, with the result that tracks frequently sank out of sight, causing derailments. See Debbie Wilkstrom, "The Horse-Drawn Street Railway: The Beginning of Public Transportation in Shreveport, 1870–1872," in *Journal of the North Louisiana Historical Association* 7 (Spring 1976): 84–7. In St. Louis, a relatively large city, streets were mostly unpaved. The four-foot-wide tracks often spread, and car wheels had to be extended on the axles. Rails would curl up, their ends sometimes penetrating the car floor. Young, "St. Louis Streetcar Story," 14, 24.

80. See Wainwright, *A Philadelphia Perspective,* 345, 521–2.

81. See, e.g., Seymour J. Mandelbaum, *Boss Tweed's New York* (New York: John Wiley, 1965), and Amy Bridges, *A City in the Republic: Antebellum New York and the Origins of Machine Politics* (New York: Cornell University Press, 1984).

82. Bridges, *City in the Republic,* 149–61; Joel A. Tarr, "The Urban Politician as Entrepreneur," *Mid-America* 49 (Jan. 1967): 55–67.

83. Gunther Barth, *City People: The Rise of Modern City Culture in Nineteenth-Century America* (New York: Oxford University Press, 1980), 110–47; Hepp, *The Middle-Class City,* 144–67.

84. Richard Harris, *Unplanned Suburbs: Toronto's American Tragedy, 1900 to 1950* (Baltimore: Johns Hopkins University Press, 1996); Robert Lewis, ed., *Manufacturing Suburbs: Building Work and Home on the Metropolitan Fringe* (Philadelphia: Temple University Press, 2004).

85. Theodore Hershberg, Harold E. Cox, Dale Light Jr., and Richard R. Greenfield, "The 'Journey-to-Work': An Empirical Investigation of Work, Residence and Transportation, Philadelphia, 1850 and 1880," in Theodore Hershberg, ed., *Philadelphia: Work, Space, Family, and Group Experience in the Nineteenth Century—Essays toward an Interdisciplinary History of the City* (New York: Oxford University Press, 1981), 4:128–73. See also Hepp, *The Middle-Class City,* 31–3; he relies on Hershberg for his data.

86. Hershberg and his coauthors calculate that streetcar fare would have taken 13% of the average unskilled worker's daily wages and almost 9% for skilled workers in both the 1850s and 1880. Hershberg et al., "The 'Journey-to-Work,'" 146–7.

87. Ibid., 141–51; Alan Gin, "Transportation Innovation and Urban Residential Location," Ph.D. diss., University of California, Santa Barbara, 1987, 120–8. See also Roger Pierce Miller, "A Time-Geographic Assessment of the Impact of Horsecar Transportation on Suburban Non-Heads-of-Household in Philadelphia, 1850–1860," Ph.D. diss., University of California, Berkeley, 1979, 288. Trips to Fairmount Park, the largest municipal park in the nation, were most frequent. See chap. 4, which deals with the influence of the horsecar on leisure patterns.

88. Harold E. Cox, "Public Transportation and the Growth of Nineteenth Century Philadelphia," paper delivered at the Eleutherian Mills Historical Library, Wilmington, Del., May 7, 1971.

89. Hershberg et al., "The 'Journey to Work,'" 151–64.

90. Richard B. Stott, *Workers in the Metropolis: Class, Ethnicity and Youth in Antebellum New York City* (Ithaca: Cornell University Press, 1990), 193–202. The figure of seventy

thousand commuters in 1855 comes from "Great Cities," *Putnam's Monthly Magazine* 5 (1855): 261, as quoted in Stott.

91. See Kenneth T. Jackson, *Crabgrass Frontier: The Suburbanization of the United States* (New York: Oxford University Press, 1985), 42–72, 105–7, for a discussion of suburbanization during the horsecar period.

92. Einhorn, *Property Rules,* 217.

93. Quoted in Jackson, *Crabgrass Frontier,* 42. An 1873 *Chicago Tribune* article noted that one of the first things a young man beginning to accumulate money did was to buy a lot in an "outlying" district near a streetcar line, build a home, and become "a member of the class of the community . . . whose business interests are in the city proper, but whose homes are away off." Quoted in Ann Durkin Keating, *Building Chicago: Suburban Developers and the Creation of a Divided Metropolis* (Columbus: Ohio State University Press, 1988), 23.

94. For Pittsburgh, see Tarr, *Changing Spatial Patterns in Pittsburgh,* 1–14; for Toronto, see Peter G. Goheen, *Victorian Toronto, 1850 to 1900: Process of Growth* (Chicago: University of Chicago Department of Geography Research Paper 127, 1970), esp. 128–33, 161–3, 177–9, 184–6, and 199–201; and for Milwaukee, see Clay McShane, *Technology and Reform: Street Railways and the Growth of a City* (Madison: State Historical Society of Wisconsin, 1974).

95. Barbara Welke, *Recasting American Liberty: Gender, Race, Law, and the Railroad Revolution, 1865–1920* (New York: Cambridge University Press, 2001), 257. Welke notes that, although in some ways "classless," streetcars were not free of "status distinctions."

96. William Dean Howells, "By Horse-car to Boston," *Suburban Sketches* (Boston: Houghton Mifflin, 1872), 91–115.

97. Clifton Hood, "Changing Perceptions of Public Space on the New York Rapid Transit System," *Journal of Urban History* 22 (Mar. 1996): 310; *NYT,* Jan. 23, 1874.

98. See "Abuse on City Cars," *NYT,* July 17, 1858; "The Rights of Passengers," *NYT,* May 20, 1871; "Courtesy in Cars, *NYT,* May 12, 1872; "The Public and the Street Cars," Jan. 23, 1874; "Hands Off," *NYT,* June 1, 1875; "Street Car Etiquette," *BE,* Feb. 25, 1876; "The Street Cars as a School of Immorality," *BE,* Mar. 20, 1876; "A Crusade against Ruffianism on the Street Cars," *BE,* Dec. 7, 1877; "A 'Masher' in Merited Misery," *BE,* Mar. 11, 1881; "Street-Car Ruffianism," *Every Saturday* 10, n.s. 2 (May 20, 1871): 458; and "The Crowded Car," *Harper's Weekly,* Sept. 21, 1872.

99. Brooks, "Street and Rapid Transit Railways," 25; Burrows and Wallace, *Gotham,* 547.

100. "Circuit Court: Elizabeth Jennings v. the Third Avenue Railroad Company," *BE,* Feb. 23, 1855. The jury awarded the plaintiff $225. See also John H. Hewitt, "The Search for Elizabeth Jennings, Heroine of a Sunday Afternoon in New York City," *New York History* 62 (1990): 387–415, and Leslie M. Harris, *In the Shadow of Slavery: African Americans in New York City, 1626–1863* (Chicago: University of Chicago Press, 2003), 270–1. In Boston, beginning in the 1840s, there had been continual challenges by abolitionists to segregated train cars. Welke, *Recasting American Liberty,* 323–31, discusses African American challenges to public transit segregation, as well as gender factors. In this case the gender of the plaintiff did not seem to play a role in the judge's decision.

101. "Colored People in City Cars," *NYT,* May 29, 1855.

102. The fact that whites could ride, if they wished, on "colored only" cars reflects the difference between these regulations and the "separate-coach" laws that emerged in the South after Reconstruction. Segregated facilities were required for blacks on some New York streetcars but were voluntary for whites. For a discussion of the implications of the separate-coach laws, see Welke, *Recasting American Liberty,* 323–75.

103. "Outrage upon a Doctor of Divinity," *NYT,* May 25, 1855, and "Card from Dr. Pennington," *NYT,* May 26, 1855. Pennington had been a slave and wrote an autobiography, *The Fugitive Blacksmith* (1849), reprinted in Arna Bontemps, ed., *Great Slave Narratives* (Boston: Beacon, 1969). The previous challenges by Pennington's followers to the segregated cars are noted in the letter to the *New York Times* by the company secretary. See also Harris, *In the Shadow of Slavery,* 271.

104. "Colored People in City Cars," *NYT,* May 29, 1855.

105. Ibid., and *NYT,* May 30, 1855. In September, a well-known African American businessman who had been ejected from one of their cars also sued. The *New York Post* ironically commented, "Better give it up, gentlemen of the Sixth-avenue. The other Railroads, more wisely, treat negroes as if they were no worse than drunken men and prostitutes, and their sense is appreciated. Your lack of it grows too apparent." Quoted in *NYT,* Sept. 2, 1855.

106. *NYT,* Dec. 19, 1856.

107. "Superior Court," *NYT,* Jan. 1, 1857; Harris, *In the Shadow of Slavery,* 271.

108. Barbara Welke reports that such exclusionary regulations "were often oral." In this case the conductor may have realized during the cross-examination that he risked putting the company in a bad light, thus endangering his job. See Welke, *Recasting American Liberty,* 326n12.

109. *NYT,* June 30, 1864. See also *BE,* Aug. 4, 1863, and *BE,* June 30, 1864. The *Eagle* supported the right of the street railroads to run cars for "colored only." Such discrimination, however, did not seem to exist on the Brooklyn streetcars, although there were some cases during the 1863 draft riots in New York City where conductors and drivers refused to allow blacks on the streetcars. See "City Railroad Accommodation," *BE,* Aug. 15, 1863, and "The Social Status of Colored People," *BE,* Jan. 2, 1865.

110. As common carriers, streetcar companies had the right to adopt "reasonable regulations," but these regulations had to be published or be known and enforced regularly. Welke, *Recasting American Liberty,* 325. The Police Commissioners heard the case because a policeman had been involved in ejecting the plaintiff.

111. The Philadelphia story is told in detail in Philip S. Foner, "The Battle to End Discrimination against Negroes on Philadelphia Streetcars: Part I. Background and Beginning of the Battle" and "Part II. The Victory," *Pennsylvania History* 40 (1973): 261–92, 355–79. See also H. E. Cox, "Jim Crow in the City of Brotherly Love: The Segregation of Philadelphia Horsecars," *Negro History Bulletin* 26 (1962): 119–23.

112. Richard Wade, *Slavery in the Cities: The South, 1820–1860* (New York: Oxford, 1964), 267.

113. There is a substantial literature on challenges to streetcar segregation. See, e.g., Catherine A. Barnes, *Journey from Jim Crow: The Desegregation of Southern Transit* (New York: Columbia University Press, 1983); Jennifer Robach, "The Political Economy of Segregation:

The Case of Segregated Streetcars," *Journal of Economic History* 46 (Dec. 1986): 893–917; and Kenneth W. Goings and Brian D. Page, "African Americans versus the Memphis Street Railway Company, Or, How to Win the Battle but Lose the War, 1890–1920," *Journal of Urban History* 30 (Jan. 2004): 131–51.

114. See, e.g., Scott Molloy, *Trolley Wars: Streetcar Workers on the Line* (Washington, D.C.: Smithsonian Institution, 1996).

115. See *NYT*, Jan. 4, 1873, for the incident involving Bergh.

Four • The Horse and Leisure

1. The phrase "frenzy for driving" comes from I. D. Ware, *The Coach-Maker's Illustrated Hand-Book* (Philadelphia, 1875), 167.

2. Thorstein Veblen, *Theory of the Leisure Class* (New York: Houghton-Mifflin, 1973, reprint of the 1899 ed.), 104–5.

3. Carolin Hodak, "Les Animaux dans la cité: Pour une histoire urbaine," *Genèses* 37 (Dec. 1999): 156–69.

4. Paul Shepard, *The Others: How Animals Made Us Human* (Washington, D.C.: Island Press, 1996), 251; Katherine C. Grier, "Animal House: Pet-keeping in Urban and Suburban Households in the Northeast, 1850–1900," in Peter Benes, ed., *New England's Creatures, 1400–1900* (Boston: Boston University Press, 1996), 109.

5. William Louis Gannon, "Carriage, Coach and Wagon: The Design and Decoration of American Horse-Drawn Vehicles," Ph.D. diss., Iowa State University, 1960, 7, 115; Ezra M. Stratton, *The World on Wheels; or Carriages, with their Historical Associations from the Earliest Time to the Present Time, Including a Selection from the American Centennial Exhibition* (New York, 1878).

6. Roy Rosenzweig and Elizabeth Blackmar, *The Park and the People: A History of Central Park* (Ithaca: Cornell University Press, 1992), 211–22, 245.

7. Ibid., 222.

8. Edwin G. Burrows and Mike Wallace, *Gotham: A History of New York City to 1898* (New York: Oxford University Press, 1999), 952–3.

9. *Brooklyn Daily Eagle* [hereafter cited as *BE*], July 17, 1875. The *Eagle* from 1841 to 1902 has been digitized and can be accessed on the Web through the Brooklyn Public Library. For a diagram of Prospect Park, see Laura Wood Roper, *FLO: A Biography of Frederick Law Olmsted* (Baltimore: Johns Hopkins University Press, 1974), 296–7.

10. Francis M. Ware, *Driving* (New York: Doubleday, Page & Co., 1903); Reginald Rives, *The Coaching Club: Its History, Records and Activities* (New York: Privately printed, 1935); James A. Garland, *The Private Stable: Its Establishment, Management, and Appointment* (Boston: Little Brown, 1899), 578.

11. Museums at Stony Brook, *Nineteenth Century American Carriages: Their Manufacture, Decoration, and Use* (Stony Brook, N.Y.: Museums at Stony Brook, 1987), 111.

12. Fairman Rogers, *A Manual of Coaching* (Philadelphia, 1910), 477.

13. William Dean Howells, *The Rise of Silas Lapham* (New York: Vintage, 1991, originally published 1885), 17.

14. Rosenzweig and Blackmar, *The Park and the People*, 219.

15. See, e.g., credit reports of Staples & Caldwell, New York City, in New York, 199:281, 300A, and Acker & Harris, New York City, in New York, 197:4, 64—R. G. Dun & Co. Collection, Baker Library, Harvard University Graduate School of Business Administration, Boston. We thank Prof. Scott Sandage, Carnegie Mellon University, for bringing these references to our attention.

16. Edward Mayhew, *Horse Doctor* (Philadelphia, 1867), 40.

17. William Louis Gannon, "Carriage, Coach and Wagon: The Design and Decoration of American Horse-Drawn Vehicles," Ph.D. diss., Iowa State University, 1960, 23, 95. See Stratton, *The World on Wheels,* 95, 446.

18. Theodore Dreiser, *Sister Carrie* (New York: Oxford University Press, 1991, originally published 1900), 286–7, 415, 458.

19. Edith Wharton, *The House of Mirth* (New York: Oxford University Press, 1994, reprint of 1905 ed.), 256; Maureen E. Montgomery, *Spectacles of Leisure in Edith Wharton's New York* (New York: Routledge, 1998), 10–13, 101; "HO! For Coney Island," *BE,* May 18, 1870; James W. Tuckerman, "Park Driving," *Outing* 46 (June 1905): 259. Some observers protested that the horse show was not "merely a social function, a parade of much-dressed men and women, with the horses in the tan-bark ring as a very thin excuse for that other exhibit in the boxes and in the promenade. . . . At the horse show it is really true that the horse is the thing." See John Gilmer Speed, "The New York Horse Show," *Leslie's Weekly* 87 (Nov. 24, 1898): 403.

20. Wharton, *The House of Mirth,* 256. Originally the horse show was intended to exhibit all kinds of horses, including workhorses, but it seems to have become much more oriented toward fashion. See, e.g., "Not an American Day" and "English Imitation," *New York Times,* Oct. 26, 1883.

21. *Rider and Driver* 1 (Mar. 7, May 16, 1891).

22. Ibid., 4 (Aug. 31, 1895).

23. Margaret Bisland, "Driving for Women," *Rider and Driver* 1 (Apr. 11, 1891): 15.

24. I. D. Ware, *The Coach-Maker's Illustrated Hand-Book* (Philadelphia, 1875), 364; *Rider and Driver* 1 (Mar. 28, 1891); James W. Tuckerman, "Park Driving," *Outing* 46 (June 1905): 260.

25. Belle Beach, *Riding and Driving for Women* (New York, 1912), 261–4.

26. Elizabeth York Miller, "Should a Woman Ride Astride," *Munsey's Magazine* 25 (July 1901): 553–7. Miller said that riding astride had grown out of the bicycling fad, and the riding academies did not like teaching it because it was more difficult and more dangerous; Maureen E. Montgomery, *Spectacles of Leisure in Edith Wharton's New York* (New York: Routledge, 1998), 101.

27. Quoted in Burrows and Wallace, *Gotham,* 720.

28. For examples, see "Horseback Riders on Bridlepath at Prospect Park, 1912" (film in the American Memory Collection of the Library of Congress, 2002). The film shows only male riders. See also Clay McShane, *Down the Asphalt Path: American Cities and the Coming of the Automobile* (New York: Columbia University Press, 1994), 33.

29. "Mardi Gras" and "Easter Parade, Fifth Ave." (films in the American Memory Collection, Library of Congress).

30. *Rider and Driver* 3 (Apr. 21, May 19, 1894).

31. "Parade of Horses on Speedway" (film in the American Memory Collection, Library of Congress).

32. C. Bennett, "At the Horse Show of 1898," *Illustrated American* 24 (Nov. 25, 1898): 412–3.

33. *BE,* Mar. 17, 1846, Dec. 21, 1879, Dec. 21, 1890, Dec. 19, 1896.

34. Blake McKelvey, *Snow in the Cities: A History of America's Urban Response* (Rochester, N.Y.: University of Rochester Press, 1995), 25, 68.

35. *BE,* Jan. 2, 1858, Jan. 31, 1871, Feb. 19, 1890. The quote is from Dec. 23, 1883.

36. *BE,* Dec. 8, 1846, Jan. 12, 1847, Jan. 2, 1850, Jan. 20, 1888, Mar. 7, 1890, Jan. 31, 1894, Dec. 14 & 18, 1896.

37. See *BE,* Dec. 14 & 18, 1896.

38. *BE,* Mar. 17, 1846, Dec. 21, 1879, Dec. 21, 1890, Dec. 19, 1896.

39. Pittsburgh, *A Digest of the Acts of Assembly Relating to and the General Ordinances of the City of Pittsburgh from 1804 to Jan. 1, 1897* (Pittsburgh: W. W. Thomson, 1897), 585; T. J. Chapman, *Old Pittsburgh Days* (Pittsburgh: J. R. Weeding, 1900), 166; Harry C. Gilchrist, *History of Wilkinsburg, Pennsylvania* (Pittsburgh, 1940), 86.

40. Melvin L. Adelman, *A Sporting Time: New York City and the Rise of Modern Athletics, 1920–1970* (Urbana: University of Illinois Press, 1986), and Stephen Reiss, *City Games: The Evolution of American Urban Society and the Rise of Sports* (Urbana: University of Illinois Press, 1979), are exceptional histories of horse racing.

41. Adelman, *A Sporting Time,* 27–40.

42. As quoted in Adelman, *A Sporting Time,* 59. Phillip Thurtle has written that the possession of trotting horses, by comparison to aristocratic thoroughbreds, was a celebration of utilitarianism. See "Breeding and Training Bastards: Distinction, Information Inheritance in Gilded Age Trotting Horse Breeding," in Phillip Thurtle and Robert Mitchell, eds., *Data Made Flesh: Embodying Information* (London: Routledge, 2004), 65–84.

43. *Boston City Council Minutes,* Jan. 13 and Apr. 7, 1890.

44. "The Charles River Speedway of Boston's Metropolitan Park System," *Engineering Record* 51 (Apr. 29, 1905): 496–8.

45. John W. Linehan and Edward Cogswell, eds., *The Driving Clubs of Greater Boston* (Boston: Atlantic Printing Co., 1914), 8–19. In this book there are photos of Curley (later the subject of the novel and film *The Last Hurrah*) and Fitzgerald (the grandfather of President John Fitzgerald Kennedy) at the reins of sulkies. Linehan and Cogswell list seven other tracks for suburban driving clubs.

46. *BE,* June 22, 1879, May, 9, 1880, Aug. 10, 1881, June 2, 1891, Mar. 29, 1892, Feb. 16, 1894, June 4, 1894.

47. *BE,* Oct. 16, 1892.

48. Ibid. Within two years of its opening, the track became embroiled in a scandal over a fixed race. See *BE,* June 2, 1894. See also *BE,* Oct. 12, 1900.

49. W. W. Stevenson et al., *The Story of the Sesqui-centennial Celebration of Pittsburgh* (Pittsburgh: R. W. Johnston Studios, 1910), 346; Barbara Judd, "Edward M. Bigelow: Creator of Pittsburgh's Arcadian Parks," *Western Pennsylvania Historical Magazine* 58 (Jan. 1975): 61; Francis G. Couvares, *The Remaking of Pittsburgh: Class and Culture in an Industrializing City, 1877–1919* (Albany: State University of New York Press, 1984), 107–11.

50. Willard Glazier, *Peculiarities of American Cities* (Philadelphia: Hubbard Brothers, 1884), 63, 119, 149, 172, 272, 301.

51. Ibid., 486. See also, on New York, Reginald Rives, *The Coaching Club: Its History, Records and Activities* (New York, 1935), 1, and *The New York Coach-Maker's Magazine*, Aug. 1860, 46. In 1861 *Carriage Monthly* reported that 467,849 "pleasure takers" in carriages had visited the park the previous year. "Carriage Roads in the Central Park," *Carriage Monthly* 4 (Nov. 1862): 195.

52. "A Plea for Better Roads," *Rider and Driver* 1 (Mar. 7, 1891): 12; Adelman, *A Sporting Time*, 55; "The Harlem River Speedway," *Engineering Record* 29 (July 1897): 350–1; "Parade of Horses on Speedway" (film in the American Memory Collection, Library of Congress); "Plans for the Speedway and Grand Concourse," *Rider and Driver* 3 (Mar. 31, 1894): 16–7; John Mullaly, *The New Parks across the Harlem* (New York: Record and Guide, 1887).

53. Scott C. Martin, *Killing Time: Leisure and Culture in Southwestern Pennsylvania, 1800–1850* (Pittsburgh: University of Pittsburgh Press, 1995), 13; Gunther Barth, *City People: The Rise of Modern City Culture in Nineteenth Century America* (New York: Oxford University Press, 1989); Patricia C. Click, *The Spirit of the Times: Amusements in Nineteenth-Century Baltimore, Norfolk, and Richmond* (Charlottesville: University Press of Virginia, 1989); Gary Cross, *A Social History of Leisure since 1600* (State College, Pa.: Venture Publishing, 1990); Karen Halttunen, *Confidence Men and Painted Women: A Study of Middle-Class Culture in America, 1830–1870* (New Haven: Yale University Press, 1982). For specific contemporary examples, see the notices for "excursions" and other such events, under the heading of "Amusements," *Brooklyn Daily Eagle* in the 1850s.

54. For cemeteries as places of recreation, see David Schuyler, *The New Urban Landscape: The Redefinition of City Form in Nineteenth-Century America* (Baltimore: Johns Hopkins University Press, 1986), 24–148, and Blanche Linden-Ward, *Silent City on a Hill: Landscapes of Memory and Boston's Mount Auburn Cemetery* (Columbus: Ohio State University Press, 1989), 209, 307–13.

55. Linden-Ward, *Silent City on a Hill*, 206–10.

56. The last set of swan boats in the country, to our knowledge, are those still operating in Boston.

57. Rosenzweig and Blackmar, *The Park and the People*, 232.

58. Linden-Ward, *Silent City on a Hill*, 313; "The Arsenal," in *The Iron City: A Compendium of Facts Concerning Pittsburgh and Vicinity, for Strangers and the Public Generally* (Pittsburgh: G. W. Pittock and K. McFall, 1867), 101; "Crowds at Fairmount," *Philadelphia North American and United States Gazette*, July 29, 1859; Burrows and Wallace, *Gotham*, 974.

59. Rosenzweig and Blackmar, *The Park and the People*, 233.

60. See, e.g., *BE*, July 7 & 9, 1855, May 30, 1861, July 3 & 4, 1863.

61. *BE*, Aug. 1, 1859.

62. *BE*, June 30, 1866, Aug. 20, 1867, Jan. 21, 1868; State of New York, *Annual Report of the State Engineer and Surveyor on the Railroads of the State* (Albany: Argus Co., 1870), 256; "To the Seaside," *BE*, Oct. 12, 1876; Brian J. Cudahy, *How We Got to Coney Island: The Development of Mass Transportation in Brooklyn and Kings County* (New York: Fordham Uni-

versity Press, 2002), 26–31, 36–123. Cudahay says that in the 1870s and 1880s the Coney Island and Brooklyn Company had problems competing with steam-powered railways on competitive routes and therefore experimented with steam-powered cars. Excursion railroads, ferries, and the elevated railroad were also major transporters of holiday seekers to Coney Island in the second half of the nineteenth century. Jon Sterngass notes that working people were "a constant presence" at Coney Island. See Jon Sterngass, *First Resorts: Pursuing Pleasure at Saratoga Springs, Newport, and Coney Island* (Baltimore: Johns Hopkins University Press, 2001), 99–100. On Coney Island history, also see John F. Kasson, *Amusing the Million: Coney Island at the Turn of the Century* (New York: Hill & Wang, 1978), and Woody Register, *The Kid of Coney Island: Fred Thompson and the Rise of American Amusements* (New York: Oxford University Press, 2001).

63. *BE,* June 14, 1870. Visitors also got to the resort via a steamer from New York City. See also *BE,* May 18, 1868, and Sterngass, *First Resorts,* 75–82.

64. Alexis McCrossen, *Holy Day, Holiday: The American Sunday* (Ithaca: Cornell University Press, 2000), 50–64.

65. In the 1860s, anti-Sabbatarians and Sabbatarians clashed over the type of activities to be permitted in Central Park. The park commissioners compromised by forbidding music and boat rentals on Sundays but permitting refreshment sales, not including beer. See Rosenzweig and Blackmar, *The Park and the People,* 254–6.

66. *BE,* Jan. 28. 1855, Jan. 31 & Feb. 7, 1856.

67. See the mayor's annual report, *BE,* Jan. 6, 1857, as well as the *BE* editorials of Jan. 3, 8, & 10, 1857.

68. *BE,* Feb. 15, Mar. 10 & 14, 1857.

69. By 1867, Sunday cars were running in not only Brooklyn but also Boston, New York, Baltimore, Washington, Philadelphia, St. Louis, and Chicago, although considerable opposition existed in most of these cities. Baltimore held a referendum on the issue in 1867, with those in favor winning 10,939–9,056. Baltimore horsecars were packed on the next two Sundays, with more than fifteen thousand people attracted to the city park where enterprising ministers had begun preaching.

Five • Stables and the Built Environment

1. *New York Times* [hereafter cited as *NYT*], May 27 & 28, 1887.

2. Ibid.

3. *NYT,* May 29, 1887.

4. *NYT,* Feb. 29, 1880, has a fascinating narrative by a reporter, describing the entire route.

5. *NYT,* May 28, 1887.

6. A census taken by the New York City Board of Health in 1896 counted 4,649 stables and 73,746 horses, or an average of approximately 16 horses per stable. This is much higher than the figures provided by the U.S. Census and very likely more accurate. See John Duffy, *A History of Public Health in New York City, 1866–1966* (New York: Russell Sage, 1974), 109.

7. Department of the Interior, Census Office, *Twelfth Census of the United States, Taken in the Year 1900, Agriculture, Part I, Farms, Live Stock and Animal Products* (1902), 321. This is actually a measure of animals "not on the farm."

8. Bureau of the Census, Department of Commerce, *Thirteenth Census of the United States, Taken in the Year 1910, Agriculture* (Washington, D.C.: Government Printing Office, 1913), 441–6.

9. The Bull's Head was a tavern frequented by horse dealers. See "At the Old Bull's Head," *Scribner's Monthly* 17 (Jan. 3, 1879): 426–32; *Report of the Council of Hygiene and Public Health of the Citizens' Association of New York* (New York: D. Appleton & Co., 1866), 81; and *NYT*, Apr. 8, 1885.

10. Mrs. O'Leary's cow is an urban legend. See Carl Smith's Web exhibit, which he curated for the Chicago Historical Society, at www.chicagohs.org/fire/intro/detail.html#oleary (visited Sept. 3, 1999).

11. Clay McShane acknowledges the labors of his seminar students James Beauchesne, Kerry Granfield, Michael Harvey, Marlene Kandall, Sandra Kelly, Chad Leinaweaver, Maureen McAuley, James McCormick, Michael O'Connor, Patrick Preston, and Sarah Swedberg in compiling this map.

12. "Street Car Service in Boston," *Street Railway Journal* [hereafter cited as *SRJ*] (June 1889): 151. *SRJ* noted a similar pattern in Paris and recommended it to its readers. *SRJ* 4 (Sept. 1889): 321.

13. Robert McClure, *Gentleman's Stable Guide* (Philadelphia, 1870), 58–9.

14. About 1870, Johnson, Shepherd & Co., proprietors of the Twenty-third Street and Ninth Avenue lines of stages, constructed a four-story brick building 75 by 100 feet at Ninth Avenue and Twentieth Street. The horses were kept in the basement and on the second floor, while the ground floor was used for storing stage coaches. The Belt Line Stable, already mentioned because of the great fire of 1887, was built in 1873. It was a three-story brick structure, extending for 200 feet along Tenth Avenue and for 420 feet on Fifty-third and Fifty-fourth streets. Horses were kept largely on the second floor and supplies on the third, with carriages on the first. *NYT*, July 29, 1873, May 28, 1887.

15. *NYT*, Dec. 3, 1876. This stable burned down in 1881.

16. *NYT*, Mar. 25, 1878.

17. *SRJ* 4 (Jan. 1888): 16; 4 (Feb. 1888): 35; 4 (July 1889): 264; 5 (Dec. 1890): 531. The Belt Line built a four-story brick stable in 1888 that would accommodate 1,600 horses on four stories, ignoring the lessons of the 1887 fire. The first floor was devoted to holding cars, with 1,200 stalls on the second floor and 400 on the third. There was an even bigger stable in Washington during the Civil War when the Army of the Potomac's remount stable in Giesboro (the site of today's Bolling Air Force Base) held better than 30,000 horses. See G. Terry Sharrer, "The Great Glanders Epizootic, 1861–66," *Agricultural History* 69 (Winter 1995): 85–6; *SRJ* 4 (Jan. 1888): 16; *SRJ* 5 (July 1889): 264; *SRJ* 6 (Dec. 1890): 531.

18. *SRJ* 4 (Jan. 1888): 16; 4 (Feb. 1888): 35.

19. *SRJ* 3 (May 23, 1887); Nancy Kenoyer, "Wells Fargo Express Horses," *Western Horseman*, Apr. 1977, 56–60; "An Expert Visits Our San Francisco Stable," *Wells Fargo Messenger* 4 (Aug. 1916): 198.

20. *NYT*, Dec. 6, 1903.

21. "Heinz Stables—Main Plant," in *H. J. Heinz Company: Producers, Manufacturers and Distributors Pure Food Products "57 Varieties"* (Pittsburgh: H. J. Heinz Co., 1910), 19. We thank Frank Kurtik, then archivist for the Heinz Family Foundation, for bringing this brochure to our attention.

22. Susan Porter Benson, *Counter Cultures: Saleswomen, Managers, and Customers in American Department Stores, 1890–1940* (Urbana: University of Illinois Press, 1988), 20.

23. One motivation for its stables' sanitary conditions and good treatment of the horses may have been to avoid the possibility of offending the store's wealthy female patrons. Conditions around the stables deteriorated, however, and in the 1890s neighbors made several complaints concerning the "filthy state" of the stables and the resulting odors. Robert W. Twyman, "History of Marshall Field and Company, 1865–1906," Ph.D. diss., University of Chicago, 1950, 340–7.

24. Ralph M. Hower, *History of Macy's of New York, 1858–1919* (Cambridge: Harvard University Press, 1943), 108, 197–8, 280–1.

25. *NYT,* Jan. 1, 1897.

26. Data for Boston from *Boston City Directory,* 1870–1900; for Pittsburgh, see *Industries of Pittsburgh: Trade, Commerce and Manufactures. Historical and Descriptive Review for 1879, and for 1880* (Pittsburgh, 1879), 80, and *Pittsburgh City Directory, 1892* (Pittsburgh, 1892). The 1892 Pittsburgh data is for livery and sale stables, and data for Philadelphia is from Philadelphia city directories.

27. *Industries and Wealth of Pittsburgh and Environs, 1890* (New York, 1890), 79, 105, 131, 146, 149, 156.

28. *Industries of Pittsburgh, Trade, Commerce and Manufactures for 1879 and 1880* (Pittsburgh, 1879); *Pittsburgh Business Directory* (Pittsburgh, 1892).

29. *Industries of Pittsburgh.*

30. See Sampson Family Papers as well as biographical sketches located in the H. John Heinz III History Center, Pittsburgh. By "mate," the biographer meant looking the equal of. Obviously geldings don't mate.

31. The organization was also known as the Livery Stablemen's Association. It served the purpose of uniting livery stable owners against the strike demands of the Liberty Dawn Association. In 1898, some of the same interests attempted to form a trust of the largest livery and cab interests in the city called the Consolidated Stable Company. However, it does not seem to have succeeded. See *NYT,* Dec. 17, 1898.

32. *NYT,* May 30 & June 1, 1881, Jan. 31, 1886.

33. This interpretation is based upon the comment in the 1880 *Pittsburgh Business Directory* (Pittsburgh, 1880) in regard to the business of R. T. Rodney, "Undertaker and Embalmer." The *Directory* commented that "Mr. Rodney was one of the first undertakers in the city to breakup the monopoly of the livery-stable keepers & bring the price of carriages to funerals down to the reasonable amount of $3.00."

34. Alexander von Hoffman, *Local Attachments: The Making of an American Urban Neighborhood, 1850 to 1920* (Baltimore: Johns Hopkins University Press, 1994), 92–4.

35. Blake McKelvey, *Snow in the Cities: A History of America's Urban Response* (Rochester: University of Rochester Press, 1995), 25–9. Some believed that sleighing was a great aphrodisiac. For more information on the use of horse-drawn sleighs, see chap. 4.

36. See "How Horses Are Hired," *Brooklyn Daily Eagle,* June 19, 1873.

37. Building permit records for Pittsburgh (located in the Archives of Industrial Society, University of Pittsburgh), first available in 1887, were examined for the decade 1890–99 for the Homewood-Brushton area, a fast-growing neighborhood on the city's eastern edge. The neighborhood had experienced rapid growth, as electric traction replaced horsecars during this period, with 829 new structures constructed from 1890 to 1899. The permit records show that fifty-five stables were constructed, although one can assume that residents probably also constructed additional ones without bothering to take out a permit. Owners built all sorts of stables, with homeowners constructing most of the smaller ones themselves. Of the 55 stables, 19 cost under $100, 17 cost $100–199, 10 cost $200–400, and 9 cost over $400. Three stables cost over $1,000. All of the stables save one were frame, with 16 having one story, 13 having a story and a half, and 24 having two stories. The small number of stables is probably explained by the limited number of horse-owning households in this middle-class neighborhood and its dependence on the electric streetcar for mobility.

38. *Report of the Council of Hygiene and Public Health of the Citizens' Association of New York* (New York: D. Appleton & Co., 1866). For a description of the council's creation and operation, see John Duffy, *A History of Public Health in New York, 1866–1966,* 2 vols. (New York: Russell Sage, 1974), 1:1–23.

39. Albert H. Buck, *A Treatise on Hygiene and Public Health* (New York: William Wood, 1871), 392.

40. *Report of the Council of Hygiene and Public Health,* 175.

41. "Number and Value of Horses in Chicago," *National Livestock Journal* 4 (1–73): 8.

42. As quoted at the Chicago Near West Side website, www.uic.edu/depts/hist/nearwest/docs/horse.html (visited Jan. 20, 2001).

43. Robert Hunter, *Tenement Conditions in Chicago* (Chicago: City Homes Association, 1901). We are indebted to Prof. Wendy Plotkin of Arizona State University at Tempe for pointing out this source. It is available on line at www.uic.edu/depts/history/northwest/ (visited Jan. 20, 2001).

44. *Report of the Council of Hygiene and Public Health,* 138.

45. Gail Harris, Landmarks Preservation Commission, "136 West 18th Street Stable," #230 LP-1818 (New York: The Commission, 1990). See also www.gvshp.org/block636a.htp and http://home.nyc.rr.com/jkn/nysonglines/18th st.htp (both visited Jan. 3, 2001).

46. James Garland, *The Private Stable: Its Establishment, Management, and Appointments* (Boston: Little Brown, 1899), 17. This book appeared in another edition the same year under the pseudonym "Jorrocks."

47. Ibid., 26–7.

48. Harris, "136 West 18th Street Stable."

49. Garland, *The Private Stable,* 59–61.

50. Ibid. The difference seems astonishingly large. Most of it represented the cost of employing a coachman/stable hand, but they were not likely to be paid $2,709 per year.

51. Robert W. Wooley, "Private Stables of Manhattan," *Outing* 38 (May 1901): 174.

52. Beacon Hill Club Stable Association Records, 1866–1914 (Massachusetts Histori-

cal Society). A few comparisons with the first, less detailed ledger of 1867 show that, while 28% of the budget went for food in both years, labor costs nearly doubled. Horses were away one-fourth of the time in 1867 but one-third in 1912, suggesting that owners were spending more time in the country or the suburbs.

53. Herbert R. Collins, "The White House Stables and Garages," *Records of the Columbia Historical Society* (1993): 366–85.

54. Francis A. Walker, *Statistics of Population at the Ninth Census: 1870, Part 1* (Washington, D.C.: Government Printing Office, 1873); *Statistics of Population at the Tenth Census: June 1, 1880* (Washington, D.C.: Government Printing Office, 1882); *Report on the Population of the United States at the Eleventh Census: 1890* (Washington, D.C.: Government Printing Office, 1897); *Thirteenth Census of the United States, Taken in the Year 1910*, Vol. 4, *Population: Occupation Statistics* (Washington, D.C.: Government Printing Office, 1902). Scott Molloy, *Trolley Wars: Streetcar Workers on the Line* (Washington, D.C.: Smithsonian Institution Press, 1996), 116, notes that the street railway company in Providence, R.I., assigned black hostlers to black teamsters.

55. Boston Work Horse Parade Association, *Seventh Annual Report* (1909), 5.

56. For a description by an insider of one such fight in a modern urban stable, see Roland L. Freeman, *The Arabbers of Baltimore* (Centreville, Md.: Tidewater Publishers, 1989), 25. Freeman, a Baltimore peddler in his youth, reported that stable hands "worked hard and partied hard." They also were stigmatized by their neighbors.

57. Edward Mayhew, *Illustrated Horse Management* (Philadelphia, 1867), 327.

58. Garland, *The Private Stable*, 313–27.

59. *NYT*, Dec. 3, 1876.

60. *NYT*, Mar. 20, 1879.

61. Albert H. Buck, *A Treatise on Hygiene and Public Health* (New York: William Wood, 1871), 392.

62. *NYT*, Feb. 13, 1877.

63. *NYT*, May 7 & 9, June 17 & 24, 1897. See also *NYT*, June 21, 1897, for protests of residents of West Eightieth Street over the conversion by the Boulevard and Amsterdam Avenue Street Railway of a vacant lot into an "open-air stable."

64. George Waring Jr., *Street Cleaning and the Disposal of a City's Wastes* (New York, 1897), n.p.

65. Naomi Rogers, "Germs with Legs: Flies, Disease, and the New Public Health," *Bulletin of the History of Medicine* 63 (1989): 599–617.

66. Carl W. Gay, *Productive Horse Husbandry* (Philadelphia: J. Lippincott & Co., 1914), 238. At least one stable keeper fed his horses molasses, a laxative, to encourage profitable defecation in the stable.

67. Von Hoffman, *Local Attachments*, 109. Nobody was caught, so the fire may have been an accident. It is a measure of the unpopularity of stables that blame seekers pointed at neighbors. Neighbors even objected to having horse-stable windows facing them. Robert McClure, *The Gentleman's Stable Guide* (Philadelphia: Porter & Coates, 1870), 20–1, noted that many persons living opposite stable windows had successfully pursued suits at law against stable owners. McClure advised the use of a window device that provided air and some light without permitting the horse to be seen or smelled from the outside.

68. *NYT,* Mar. 15, 1871. In 1848, the *Brooklyn Daily Eagle* reported that, after a fire in an omnibus stable that killed 150 horses who "perished most miserably in the flames, the burnt carcasses of these poor animals lay strewed in heaps over the ground." See *Brooklyn Daily Eagle,* May 26 & 27, 1848.

69. See, e.g., *NYT,* July 4, 1866, Mar. 15, 1871, Dec. 5, 1872, June 13, 1886.

70. *NYT,* Dec. 3. 1876.

71. *NYT,* Dec. 5, 1872, June 25, 1874, Oct. 11 & 12, 1881, July 22 & 24, 1889.

72. *NYT,* Oct. 11 & 18, 1881.

73. *NYT,* May 29, 1887.

74. *NYT,* July 22 & 24, 1889.

75. *Common Council Minutes,* Mar. 22 & Apr. 21, 1881. No neighborhood opposition surfaced, although the neighborhood's alderman reported fears that the stable would actually hold 450 horses. We can only speculate that South Boston's blue-collar residents were willing to put up with the nuisance in return for the steady jobs that the stables generated.

76. *Sanborn Fire Insurance Map of Boston,* 1885.

77. Unless otherwise noted, information in this and succeeding paragraphs comes from the *Annual Report of the New York City Board of Health* on the dates noted in the text. In 1913, Philadelphia required that manure pits drain into sewers. An 1895 ordinance banned manure pits under sidewalks or in front of premises, while a 1913 ordinance required that manure be stored in a water- and fly-proof bin not closer than two hundred yards from churches, dwellings, and schools. See Philadelphia, *Rules and Regulations for the Sanitary Keeping of Stables* (Philadelphia, 1913).

78. Robert McClure, ed., *Every Horse Owner's Cyclopedia* (Philadelphia, 1871), 155.

79. Albert H. Buck, *A Treatise on Hygiene and Public Health* (New York: William Wood, 1871), 155.

80. "No Place to Dump Stable Manure—What Is to Become of It?" *NYT,* July 21, 1881. There are several possible explanations. The supply may have exceeded the demand as urban herds grew. Possibly farmers were substituting other fertilizers, perhaps guano from Chile. In the *New York Times* of July 11, 1888, stable keepers complained that they had to pay one dollar a cartload for manure removal. Previously they had received fifty cents a load. Manure was piling up in the stables because keepers were boycotting the manure contractors.

81. *NYT,* Nov. 18, 1884.

82. *NYT,* Dec. 10 & 17, 1884.

83. *NYT,* Dec. 10, 17, 19, & 23, 1884.

84. *NYT,* Dec. 24, 1884. The last complaint appeared in the *NYT* in 1888.

85. *Annual Report of the New York City Board of Health* (1915), 12.

86. Robert Hunter, *Tenement Conditions in Chicago* (Chicago: City Homes Association, 1901); Edith Abbott, *The Tenements of Chicago, 1908–1935* (Chicago: University of Chicago Press, 1936), 477–8.

87. In *The Tenements of Chicago,* 477–8, Abbott observes that, "Along with the disappearance of the horse from the Chicago streets went the filthy stables and the dreadful manure heaps that accumulated in the alleys of the tenement districts."

88. See, e.g., "The Alleys of New York," for Brooklyn and Manhattan at www.forgotten-ny.com/Alleys/ALLEYS%20HOME/alleys.html (visited Jan. 3, 2004).

Six • Nutrition

1. Stephen Budiansky, *The Nature of Horses: Exploring Equine Evolution, Intelligence and Behavior* (New York: Free Press, 1997), 15; W. H. Jordan, "The Feeding of Animals," in L. H. Bailey, ed., *Cyclopedia of American Agriculture: A Popular Survey of Agricultural Conditions, Practices and Ideals in the United States and Canada*, 4 vols. (New York: Macmillan, 1908), 3:78–87.

2. Horse Association of America, "Grain Surplus Due to Decline in Horses," *Bulletin 199*, n.d., and "Shifting Uses of Land," *Bulletin 209*, n.d., in National Agricultural Library, Beltsville, Md. In 1918 the National Automobile Chamber of Commerce reported that 10 million acres of land "heretofore required to sustain horses [had been] released by motor truck for production of human foods," as noted in *Forty-fifth Annual Report of the Western Pennsylvania Humane Society* (Pittsburgh, 1919), 35. One agricultural historian has estimated that, in 1910, 16 million acres out of 325 million acres of harvested cropland were used to feed urban horses, while another 72 million acres were devoted to providing food for farm horses and mules. See William L. Cavert, "The Technological Revolution in Agriculture, 1910–1955," *Agricultural History* 30 (Jan. 1956): 19–20. The history of the Horse Association of America is discussed in detail in Alan L. Olmstead and Paul W. Rhode, "The Agricultural Mechanization Controversy of the Interwar Years," *Agricultural History* 68 (Summer 1994): 35–53. Olmstead and Rhode discuss the effects of the decline of the horse on American society, arguing that it caused sweeping changes in both city and farm.

3. For discussion of the von Thunen model in relationship to American cities and towns, see Cavert, "The Technological Revolution in Agriculture"; Diane Lindstrom, *Economic Development in the Philadelphia Region, 1810–1850* (New York: Columbia University Press, 1978); Timothy R. Mahoney, *River Towns in the Great West: The Structure of Provincial Urbanization in the American Midwest, 1820–1870* (New York: Cambridge University Press, 1990), 178; and E. Melanie Dupuis, "Making the Country Work for the City: Von Thunen's Ideas in Geography, Agricultural Economics and the Sociology of Agriculture," *American Journal of Economics and Sociology* 60 (Jan. 2001): 1–15.

4. Winifred Barr Rothenberg, *From Market-Places to a Market Economy: The Transformation of Rural Massachusetts, 1750–1850* (Chicago: University of Chicago Press, 1992); Donald H. Parkerson, *The Agricultural Transition in New York State: Markets and Migration in Mid-Nineteenth-Century America* (Ames: Iowa State University Press, 1995), 79–102.

5. John R. Quinn, *Fields of Sun and Grass: An Artist's Journal of the New Jersey Meadowlands* (New Brunswick, N.J.: Rutgers University Press, 1992), 85–8. Some farmers preferred saltwater hay to supplement the horse diet with salt.

6. Gordon G. Whitney, *From Coastal Wilderness to Fruited Plain: A History of Environmental Change in Temperate North America, 1500 to the Present* (New York: Cambridge University Press, 1994), 250–5; Daniel Vickers, *Farmers and Fishermen: Two Centuries of Work in Essex County, Massachusetts, 1630–1850* (Chapel Hill: University of North Carolina Press, 1994), 295–6.

7. Quinn, *Fields of Sun and Grass,* 85–8.

8. Willard W. Cochrane, *The Development of American Agriculture: A Historical Analysis,* 2d ed. (Minneapolis: University of Minnesota Press, 1993), 73–7.

9. Vickers, *Farmers and Fishermen,* 295–6.

10. Howard S. Russell, *A Long, Deep Furrow: Three Centuries of Farming in New England* (Hanover, N.H.: University Press of New England, 1976), 367.

11. Ronald Dale Karr, "The Transformation of Agriculture in Brookline, 1770–1885," *Historical Journal of Massachusetts* 15 (Jan. 1987): 33–9. A similar pattern existed in Concord, also in the Boston hinterland. Hay production, both meadow hay and English hay, increased from 11.4 tons per producer in 1801 to 27.7 tons per producer in 1850. See Robert Gross, "Culture and Cultivation: Agriculture and Society in Thoreau's Concord," *Journal of American History* 69 (June 1982): 60.

12. *Census of Massachusetts: 1875. Compendium,* 23. The salt hay industry had almost totally disappeared by 1895, not just because the price of upland hay had declined but also because the city was filling in the marshes for residential development.

13. Lindstrom, *Economic Development in Philadelphia,* 140–5, 182.

14. Parkerson, *Agricultural Transition in New York State,* 90–102.

15. D. J. Browne, *The American Muck Book: Treating of the Sources, History, and Operations of All the Principal Fertilizers and Manures* (New York: C. M. Saxton, 1851); see also Mark Linder and Lawrence S. Zacharias, *Of Cabbages and Kings County: Agriculture and the Formation of Modern Brooklyn* (Iowa City: University of Iowa Press, 1999), 49.

16. Browne, *The American Muck Book,* 250; Linder and Zacharias, *Of Cabbages and Kings County,* 33, 39–41, 53. Richard A. Wines, *Fertilizer in America* (Philadelphia: Temple University Press, 1985), 10, 259, estimates that urban manures had one-twentieth the value of modern fertilizers.

17. Jimmy Skaggs, *The Great Guano Rush: Entrepreneurs and American Overseas Expansion* (New York: St. Martin's Griffin, 1994), 5–144. Guanos were mostly used outside urban hinterlands, as in the rural South.

18. Terence Young, "From Manure to Steam: The Transformation of Greenhouse Heating in the United States, 1870–1900," *Agricultural History* 72 (Summer 1998): 574–97. Steam ultimately displaced manure as a heating agent, largely because it produced more predictable temperatures.

19. Joel A. Tarr, "From City to Farm: Urban Wastes and the American Farmer," *Agricultural History* 49 (Oct. 1975): 602–5; Wines, *Fertilizer in America,* 8–20. The 1866 Citizens' Association *Report upon the Sanitary Condition of the City,* cxxvi, noted Liebig's theory about using urban wastes to fertilize the land.

20. John R. Stilgoe, *The Common Landscape of America, 1580 to 1845* (New Haven: Yale University Press, 1982), 187.

21. Paul W. Gates, *The Farmer's Age: Agriculture, 1815–1860,* vol. 3 of *The Economic History of the United States* (New York: Holt, Rinehart & Winston, 1960), 250; Allen R. Yale Jr., *While the Sun Shines: Making Hay in Vermont, 1789–1990* (Montpelier: Vermont Historical Society, 1991), 6–15.

22. Clarence H. Danhof, "Gathering the Grass," *Agricultural History* 30 (Oct. 1956), 169; Percy Wells Bidwell and John I. Falconer, *History of Agriculture in the Northern United*

States, 1620–1860 (Washington, D.C.: Carnegie Institution, 1925), 296–7, 372; Parkerson, *Agricultural Transition in New York State*, 91–2.

23. For comments about the importance of terrain in the adoption of agricultural improvements, see Paul A. David, "The Landscape and the Machine: Technical Interrelatedness, Land Tenure and the Mechanization of the Corn Harvest in Victorian Britain," in David, *Technical Choice, Innovation and Economic Growth: Essays on American and British Experience in the Nineteenth Century* (New York: Cambridge University Press, 1975), 237–41; Clarence H. Danhof, *Change in Agriculture: The Northern United States, 1820–1879* (Cambridge: Harvard University Press, 1969), 220, 232, 241.

24. Quoted in Danhof, "Gathering the Grass," 172–3; R. Douglas Hurt, *American Farm Tools from Hand-Power to Steam-Power* (Manhattan, Kans.: Sunflower University Press, 1982), 85.

25. When the Middle West and the Great Plains became great hay-growing areas in the late nineteenth century, side-delivery rakes and sweep-rakes, both of which operated well on the level prairies, took over. Hurt, *American Farm Tools*, 86–7.

26. Danhof, "Gathering the Grass," 231–3; Danhof, *Change in Agriculture*, 228–49; Robert Leslie Jones, *History of Agriculture in Ohio to 1880* (Kent, Ohio: Kent State University Press, 1983), 274–5. A further step after mowing the hay was to fluff it with a "tedder" so it would dry properly before storing. See Hurt, *American Farm Tools*, 91. McCormick, the largest of the reaper makers, overcame some of these objections by becoming one of the first firms to extend consumer credit.

27. Hurt, *American Farm Tools*, 92–9; Gates, *The Farmer's Age: Agriculture*, 251.

28. Harry B. McClure, *Market Hay*, Farmers' Bulletin 508, U.S. Department of Agriculture (Washington, D.C.: Government Printing Office, 1912).

29. Rothenberg, *From Market-Places to a Market Economy*, 85–95. This process was frequently followed with regard to meadow hay, which was extremely bulky.

30. David E. Schob, *Hired Hands and Plowboys: Farm Labor in the Midwest, 1815–1860* (Urbana: University of Illinois Press, 1975), 43–58.

31. Albert Todd, "Hay Trade from a Shipper's Standpoint," *Report of the Seventh Annual Meeting of the National Hay Association* (Chicago: Geo. W. Ingersoll, 1900), 188–9.

32. Russell, *A Long, Deep Furrow*, 131–2.

33. George S. Blakeslee, "The National Hay Association," *Report of the Sixth Annual Meeting of the National Hay Association Held at Detroit, Michigan, Aug. 8, 9, 10, 1899* (Cincinnati: Earhart & Richardson, 1899), 57.

34. Quoted in Gates, *The Farmer's Age: Agriculture*, 252. A thirty-foot-high hay press built in 1849–50 still stands in a barn in Leavenworth, Indiana. The press produced bales weighing 280–400 pounds, and the compressed hay weighed about 12.6 pounds per cubic foot. The heavy bales were pushed down a chute into a hay boat on the Ohio River. See Warren E. Roberts, "An Early Hay Press and Barn on the Ohio River," *Material Culture* 25 (1993): 29–35.

35. Jones, *History of Agriculture in Ohio*, 276–7.

36. Nourse, Mason & Co., *Descriptive and Illustrated Catalogue of Plows and Other Agricultural and Horticultural Implements and Machines* (Worcester, Mass., 1857), 14–5; C. V. Piper, *Growing Hay in the South for Market, Farmers' Bulletin 677*, U.S. Department of Agri-

culture (Washington, D.C.: Government Printing Office, 1915), 16. The hay press was similar to the cotton press and was sometimes used for both commodities.

37. McClure, *Market Hay*, 17–22.

38. Piper, *Growing Hay in the South*, 17.

39. Jones, *History of Agriculture in Ohio*, 278; Cyrus H. Bates, "Business Methods Regarding Shippers and Receivers," *Report of the Seventh Annual Meeting of the National Hay Association, 1900*, 68–75; U.S. Department of Agriculture, *High-Grade Alfalfa Hay: Methods of Producing, Baling, and Loading for Market*, Bulletin 1539 (Washington, D.C.: Government Printing Office, 1903), 21–4.

40. A. Patriarche, "The Transportation of Hay," *Report of the Seventh Annual Meeting of the National Hay Association, 1900*, 172–5. The Paris omnibus company began to bale its hay in 1877. The company reported that it had reduced the bulk of hay by two-thirds, saving 50% in transport costs. Ghislaine Bouchet, *Le Cheval à Paris de 1850 à 1914* (Geneva: Libraire Droz, 1993), 106.

41. "Hay Statistics," *Report of the Seventh Annual Meeting of the National Hay Association, 1900*, 48. In 1912 a Department of Agriculture hay specialist estimated that farmers exported 22% of the 1908 hay crop (more than 15 million tons) to urban markets. See McClure, *Market Hay*, 5.

42. U.S. Bureau of the Census, "Hays and Forage," in *Twelfth Census of the United States* (1900), 6:201–8.

43. In New Hampshire from 1880 to 1900, hay production increased by 11% and revenues from hay by 15%. Some of the increased hay production went to feed farm livestock, but a substantial amount was sold to urban markets. See Paul Glenn Munyon, "A Critical Review of Estimates of Net Income from Agriculture for 1880 and 1900: New Hampshire, a Case Study," *Journal of Economic History* 37 (Sept. 1976): 643–4, 653.

44. *Census of Massachusetts: 1885*, vol. 3, *Agriculture; Census of Massachusetts: 1905*, vol. 3, *Agriculture*.

45. The 1888 western hay crop was the largest ever raised, while eastern and southern crops were average. The large crop drove down prices and caused much substitution of upland prairie hay for timothy hay, grown in the East. Chicago merchants benefited especially. See Chicago Board of Trade [hereafter cited as CBT], *Thirty-first Annual Report*, Dec. 31, 1888, xli–xlii.

46. James Morris Berry, "Marketing Oats in the United States," B.S. thesis, University of Cincinnati, 1926, 4.

47. Ibid., 55–6.

48. Jeremy Atack and Fred Bateman, *To Their Own Soil: Agriculture in the Antebellum North* (Ames: Iowa State University Press, 1987), 172; William Cronon, *Nature's Metropolis: Chicago and the Great West* (New York: W. W. Norton, 1991), 100.

49. Berry, "Marketing Oats in the United States," Table V: Farm Stocks, Crop, Farm Consumption, and Shipments . . . of Total Supplies of Oats in the United States, 1897–1925.

50. Hurt, *American Farm Tools*, 24–30, 40–59; Danhof, *Change in Agriculture*, 206–49.

51. Hurt, *American Farm Tools*, 69–76; Danhof, *Change in Agriculture*, 221–7.

52. Cronon, *Nature's Metropolis,* 104–9.

53. Jonathan Lurie, *The Chicago Board of Trade, 1859–1905: The Dynamics of Self-regulation* (Urbana: University of Illinois Press, 1979), 26–7; Cronon, *Nature's Metropolis,* 116–20; Richard Edwards, ed., *Origin, Growth and Usefulness of the New York Produce Exchange* (New York: Historical Publishing Co., 1884), 45–6. Before the Produce Exchange was founded in 1862, produce merchants transacted their business in the Corn Exchange.

54. Mixing was a process whereby processors diluted higher quality grain with lower until quality dropped to the minimum standard for that grade level.

55. CBT, *Twenty-sixth Annual Report, 1883,* 16; CBT, *Twenty-seventh Annual Report, 1884,* xxii; Berry, "Marketing Oats in the United States," 38, 59–61, 78–9.

56. Rothenberg, *From Market-Places to a Market Economy,* 91.

57. See Chicago, *Laws and Ordinances Governing the City of Chicago,* Jan. 1, 1866 (Chicago, 1866), 249–50; Daniel J. Sweeney, comp., *Ordinances of the City of Buffalo* (Buffalo, 1920), 224; *The Charter and Ordinances of the City of Battle Creek* (Battle Creek, Mich., 1861), 105; *Revised Charter and Ordinances of the City of Detroit* (Detroit, 1855), 182; and *Revised Ordinances of the City of Galesburg* (Galesburg, Ill., 1863), 154. A good description of a municipal hay market can be found in Myron Tuthill Bly, "The Hay Market," in Edward R. Foreman, comp. and ed., *Centennial History of Rochester, New York,* vol. 3, *Expansion* (Rochester, 1933), 127–9. We thank Diane Shaw for this reference. The Cleveland hay market is described in *The Encyclopedia of Cleveland History,* accessed on line at http://ech.cwru.edu/ech-cgi/article.pl?id=H.

58. See *Ordinances Governing the City of Chicago,* 1866, 248–50; *Ordinances of the City of Battle Creek,* 1861, 101–4; *Ordinances of the City of Detroit,* 1855, 182; *Ordinances of the City of Galesburg,* 1863, 52–153; and W. W. Thomson, comp., *A Digest of the Acts of Assembly Relating to, and the General Ordinances of the City of Pittsburgh from 1804 to Sept. 1, 1886* (Harrisburg, 1887), 353.

59. CBT, *Twentieth Annual Report, 1878,* 16; Edwards, *New York Produce Exchange,* 49. New York standards, however, were somewhat different from those of the National Hay Association, esp. in regard to clover hay. See comments of J. D. Carscallen in *Seventh Annual Meeting of the National Hay Association,* 115–7.

60. Todd, "Hay Trade from a Shipper's Standpoint," 188–92; James E. Boyle, *Marketing of Agricultural Products* (New York: McGraw-Hill, 1925), 308–9.

61. "The Hay Trade," in *Report of the Industrial Commission on the Distribution of Farm Products* (Washington, D.C.: Government Printing Office, 1901), 6:417; Boyle, *Marketing of Agricultural Products,* 309. In some cases, shippers surrounded bales with wooden slats to increase their weight for sale. In 1875, New York City produce merchants persuaded the New York City Common Council to approve an ordinance limiting the weight of wood frames to five pounds. *New York Times* [hereafter cited as *NYT*], Mar. 18, 1875. In 1882, the *Times* reported that there were still complaints about the use of "heavy wood" in baling (Apr. 22, 1882).

62. *Constitution and By-laws: The National Hay Association, Inc.* (St. Petersburgh, Fla.: National Hay Association, n.d.), 1. See also Todd, "Hay Trade from a Shipper's Standpoint," 190–1.

63. *NYT,* Apr. 22, 1882.

64. "Grades of Hay," *NYT,* Sept. 13, 1889. The New York State Hay and Straw Dealers' Association had established the standards.

65. CBT, *Thirty-fifth Annual Report,* 1892, xl; Todd, "Hay Trade from a Shipper's Standpoint," 188–92; Boyle, *Marketing of Agricultural Products,* 308–9; *Constitution and By-laws: National Hay Association,* 1; "The Hay Trade," 415.

66. The standards of the New York Produce Exchange were similar to those of Chicago, but they also graded and inspected "straw." New York Produce Exchange, *Report from July 1, 1896 to July 1, 1897* (New York: Jones Printing Co., 1997), 317–20.

67. CBT, *Twentieth Annual Report of the Trade and Commerce of Chicago for the Year Ending Dec. 31, 1877* (Chicago: Knight & Leonard, 1878), lxxxiii–lxxxiv.

68. CBT, *Twenty-seventh Annual Report, 1884,* xxxvi.

69. Glenn Porter and Harold C. Livesay, *Merchants and Manufacturers: Studies in the Changing Structure of Nineteenth-Century Marketing* (Baltimore: Johns Hopkins Press, 1971), 198, 214–5.

70. Harry B. McClure, *Conditions Affecting the Value of Market Hay,* U.S. Dept. of Agriculture Farmers' Bulletin 362 (Washington, D.C.: Government Printing Office, 1909), 23–30.

71. *Constitution and By-laws: National Hay Association,* 2, 12–3, 23–4; CBT, *Forty-second Annual Report* (Dec. 31, 1899), lxi; McClure, *Market Hay,* 23–4.

72. Richard L. Kohls, *Marketing of Agricultural Products* (New York: Macmillan Co., 1955), 24–6; "National Hay Association vs. L.D.&M.S.R.R. et al.: Report of the Interstate Commerce Commission," in *Report of the Ninth Annual Convention,* National Hay Association (Chicago, 1902), 133–42.

73. Kohls, *Marketing of Agricultural Products,* 25. See also McClure, *Market Hay,* 21–3, 31–2.

74. J. H. S. Johnstone, *The Horse Book: A Practical Treatise on the American Horse Breeding Industry as Allied to the Farm* (Chicago: J. H. Sanders Publ. Co., 1908), 56.

75. McClure, *Market Hay,* 7; Michael A. Balas and John E. Bayor, eds., *New Holland Haymaker's Handbook* (New Holland, Pa.: New Holland, 1987); John M. Kelly, *Handbook of Greater Pittsburg* [*sic*] (Pittsburgh: J. M. Kelly Co., 1895), 38–9.

76. McClure, *Market Hay,* 25–36; Balas and Bayor, *Haymaker's Handbook,* 174.

77. *NYT,* Mar. 4, 1900.

78. George H. Thurston, *Pittsburgh as It Is: Facts and Figures, Exhibiting the Past and Present of Pittsburgh—Its Advantages, Resources, Manufactures, and Commerce* (Pittsburgh: W. S. Haven, 1857), 183.

79. *McElroy's Philadelphia [and] Pennsylvania City and Business Directory for 1859* (Philadelphia: E. C. & J. Biddle, 1859).

80. James Garland, *The Private Stable: Its Establishment, Management, and Appointments* (Boston: Little Brown, 1899), 391–401; J. A. Simms and J. O. Williams, *Hay Requirements of City Work Horses,* Bulletin 173 (Storrs: Connecticut Agricultural College, May 1931), 8–9.

81. A New York City cartman active from the 1840s to the 1870s mentions leading his horse to its stable one evening, taking "a shilling bundle of hay out of the feed-store over

the stable," and paying for it at the "grocery on the corner." See Isaac S. Lyon, *Recollections of an Old Cartman* (New York: New York Bound, 1984, reprint of 1872 ed.); American Society for the Prevention of Cruelty to Animals (ASPCA) archives, file 1. For fountains, see the records at the society's headquarters in Manhattan. For other cities, see "Drink for Man and Beast," *NYT,* May 14, 1869 (Brooklyn); "Water to Drink," *Brooklyn Daily Eagle* [hereafter cited as *BE*], July 10, 1873 (Brooklyn); "New Drinking Fountain," *BE,* June 18, 1898; "To Build More Fountains," *BE,* June 22, 1901; and *Philadelphia and Its Environs* (Philadelphia: Lippincott, 1873), 36–7. A few ornamental troughs/fountains remain, often used as planters. For concern over fountains as a source of disease and prohibitions against their use, see *NYT,* Sept. 11, 1872 (summarizing the *Philadelphia Press*); *Pittsburgh Ordinances, 1887* (Pittsburgh, 1887), 432; and Wells Fargo & Co. Express, *Rules and Instructions* (San Francisco, 1914), leaflet at the Wells Fargo archives, San Francisco.

82. For examples of underfeeding, see "Grain Feeding of Young Horses vs. Starving as a Means of Toughening," *Annual Report of the State Board of Agriculture of Ohio* (1891), 151–4. "Statistical Report," *Annual Report of the State Board of Agriculture of Ohio* (1891), 38, provides mortality data by county. Urban/rural comparisons must be used with great care, since urban employers did not buy horses until they were five years old and often sold older ones back to farmers.

83. John Stewart, *The Stable Book* (New York: A. O. Moore & Co., 1859), 225.

84. Newton Rector, "Where the Draft Horse Excels and Pays," *Annual Report of the State Board of Agriculture of Ohio* (1891), 134–50, 152; J. H. S. Johnstone, *The Horse Book: A Practical Treatise on the American Horse Breeding Industry as Allied to the Farm* (Chicago: Sanders Publ. Co., 1908), 56; *Street Railway Journal* 2 (May 1888): 245.

85. Gerald Carson, *Men, Beasts, and Gods: A History of Cruelty and Kindness to Animals* (New York: Scribner's, 1967), 77, gives examples. The articles in the ASPCA clipping files in their New York City archives suggest more complaints about malnourished horses in the late winter, mostly in small stables.

86. Anson Rabinbach, *The Human Motor: Energy, Fatigue and the Origins of Modernity* (Berkeley and Los Angeles: University of California Press, 1990), 3.

87. Justus von Liebig, *Animal Chemistry* (Philadelphia: Campbell, 1843), 3; Rabinbach, *The Human Motor,* 65.

88. In 1850, one of Liebig's students published a table giving the following equivalents for 100 pounds of wheat: oats, 117; corn (maize), 138; carrots, 95; and turnips, 133—figures that numerous scientists, including Harvard chemist Eben Horsford, refined in the next ten years. See Elmer Verner McCollum, *A History of Nutrition* (Boston: Houghton-Mifflin, 1957), 95.

89. "Feeding Animals—Horses," *National Livestock Journal* 11 (Dec. 1880): 508–10; J. Stonehenge, "Theory and Practice of Feeding" (originally published in an unnamed British journal), *Annual Report of the State Board of Agriculture of Ohio* (1863), 182–96; Robert Leslie Jones, "The Horse and Mule Industry in Ohio to 1865," *Mississippi Valley Historical Review* (Mar. 1933): 65–8; "Stables and the Care of Horses," *American Railroad Journal* (Dec. 1884): 305–6; R. Kay, "Railroad Horses: Their Selection, Management, Some of Their Diseases and Treatment," *American Veterinary Review* (Feb. 1885): 207–13; Ghislaine Bouchet, *Le Cheval à Paris de 1850 à 1914* (Geneve: Libraire Droz, 1993), 105–37; Nicholas Papayanis,

The Coachmen of Twentieth Century Paris: Service Workers and Class Consciousness (Baton Rouge: LSU Press, 1993), 2. Occasionally, wheat prices fell to the point that the grain was substituted. See G. E. Morrow, "Economical Feeding," in Ohio State Agricultural Board, *Annual Report* (1894), 567.

90. Stewart, *The Stable Book,* 183.

91. Ibid., 253.

92. Jordan, "The Feeding of Animals," 3:56–103.

93. E. F. Bowditch, "Percheron Horses," *Massachusetts Board of Agriculture Annual Report* 34 (1886): 101. At least one firm immediately sold newly bought horses if they were not good feeders. Caloric consumption figures were available in such detail that owners knew that 300 calories were required for each 500 pounds of horse per level mile and that 34% more was needed if the horse was climbing an 11% grade; other increments were needed when the weight pulled grew by a ton.

94. Stewart, *The Stable Book,* 225; G. W. Butler, "Hygiene and Its Relation to Horse," *Fifth Annual Report of the Ohio State Livestock Commission* (1893): 427.

95. Bert W. Bierer, *History of the Animal Plagues of North America with Occasional References to Diseases and Other Conditions* (USDA, 1974 reprint—no information on original publication).

96. Dr. Peterson et al., *Special Report on Diseases of the Horse* (Washington, D.C.: U.S. Department of Agriculture, 1947), 2.

97. *Street Railway Journal* 4 (Apr. 1888): 99.

Seven • Health

1. H. H. Cole and W. N. Garrett, *Animal Agriculture: The Biology, Husbandry and Use of Domestic Animals* (New York: W. H. Freeman & Co., 1980), 708.

2. Organizing the data is complex. The graphs use raw deaths for horses and humans, not rates, because there is no accurate data on equine populations. Although the numbers of deaths for both species were increasing, both populations were growing at a rapid pace. In other words, mortality was up, but mortality rates were probably down. For comparability we adjusted the data to eliminate infant mortality among humans, since horses rarely moved to cities before the age of five. Infant mortality, especially from diarrheal diseases, was common among humans. Nor was it possible to get precise data on the causes of death for horses. Deaths for the two species, compared on a monthly or an annual basis, correlated highly statistically. The "spiky" nature of both graphs suggests that death came from mostly epidemic diseases for humans and epizootic diseases for horses. This suggests that horses and humans suffered from the same diseases or that the same environmental changes (such as heat waves or snow) depressed both species' immune systems at the same time, opening the way for infections. There is also some possibility of zoonoses, diseases that pass from species to species.

3. S. W. Tromp, *Medical Biometeorology: Weather, Climate and the Living Organism* (Amsterdam: Elsevier Publishing Co., 1963); G. M. Morton and Joan Morton, "Environment, Temperature, and Death Rates," *Age and Aging* 7 (1978): 210–24. Both the Mortons and Tromp were writing about humans, but their rationale should work for horses also.

4. S. S. Field, "Mortality of Horses in New York," *American Veterinary Review* [hereafter cited as *AVR*] 10 (July 1886): 274–5; W. L. Williams, "The Influence of Climate and Other Environments on the Distribution and Character of Disease," *AVR* 18 (Nov. 1894): 542 *et seq*. Williams notes an increase for other cities but does not have New York–specific data.

5. *New York World*, Sept. 23 & Oct. 26, 1886.

6. *New York Times* [hereafter cited as *NYT*], May 27, 28, 29 & June 1, 1887.

7. As reported in "Heat Deaths in New York," *Street Railway Journal* 2 (Aug. 1887): 669. There are no data to allow comparison with other years.

8. *ASPCA Annual Report*, 1897. Despite its name, the American Society for the Prevention of Cruelty to Animals was purely a New York organization at the start. The most common reason for the ASPCA ordering a horse off the streets was sores. Urban horses suffered from several skin infections, notably mange, and rubbing an infected spot with a harness would lead to sores.

9. Gerald Carson, *Men, Beasts, and Gods: A History of Cruelty and Kindness to Animals* (New York: Scribner's, 1967), 77; *ASPCA Annual Report* 1 (1867): 14, 27; George T. Angell, *Autobiographical Sketches and Personal Recollections* (Boston: Franklin Press, 1874), 20. *ASPCA Annual Report* 17 (1884) reported 488 abandoned horses, mostly lame.

10. Ghislaine Bouchet, *Le Cheval à Paris de 1850 à 1914* (Geneve: Libraire Droz, 1993), 92.

11. Rufus M. Steele, "Killing an Army of Horses to Rebuild San Francisco," *Harper's Weekly* 51 (Apr. 20, 1907): 580–1.

12. *NYT*, Feb. 25, 2003.

13. J. C. Meyer, "Poison in Ammonia," *AVR* 7 (Jan. 1884): 445–7. The report did not give the location of the Moerlein brewery, where the accident occurred. It was probably in the New York area, where the *Review* was published.

14. *Boston Globe*, Jan. 15, 1919. The force of the molasses flood bent pillars on the nearby elevated railroad, forcing it out of service.

15. *Annual Report of the New York City Board of Health* (1870), 237.

16. John S. Billings, *Relation of Animal Diseases to the Public Health and Their Prevention* (New York: D. Appleton & Co., 1884), 177.

17. Ibid., 122; Abram S. Benenson, *Control of Communicable Diseases in Man* (Washington, D.C.: American Public Health Association, 1990), 17, 18, 23, 25–6; Barry Dawkins and Janice Audin, *Zoonosis Updates from the Journal of the American Veterinary Association* (Schaumberg, Ill.: American Veterinary Medicine Association, 1990); Christopher Andrewes [*sic*] and John R. Walton, *Viral and Animal Zoonoses*, 17, 25–6; J. Warren Evans, Anthony Borton, Harold F. Hontz, and L. Dale Van Vleck, *The Horse* (San Francisco: W. H. Freeman & Co., 1977), 57; Abram S. Benenson, *Control of Communicable Diseases in Man* (Washington, D.C.: American Public Health Association, 1990), 73, 430; B. W. Bierer, *History of the Animal Plagues of North America* (n.p., n.d.); A. Drinkwater, "Typhoid Fever in Horses," *AVR* 2 (Feb. 11, 1878): 102. The worst outbreak of equine encephalitis, in 1938, killed 184,000 horses and 3,000 humans.

18. New York City, *Annual Report of the Board of Health* (1910), 23.

19. *Annual Reports of the State Board of Agriculture of Ohio*, 1879–1886. The graph gives a somewhat false illusion of precision. The Ohio State Board of Agriculture collected

county-level information on horse populations from the state auditor, since owners paid a personal property tax on their steeds. This probably led to an undercount because tax avoidance was commonplace. We don't know the manner of collecting information on deaths, although perhaps it came from local veterinarians. It, too, is a likely understatement, since the life expectancy for horses probably did not exceed twenty years, which would suggest an annual mortality rate of about fifty per thousand, much higher than this. Many owners apparently did not bother to report deaths, especially for foals. Union County produced many draft animals, so its horses would have been genetically similar to those in big cities.

20. See "Grain Feeding of Young Things v. Starving as a Means of Toughening," *Annual Report of the State Board of Agriculture of Ohio* (1891), 492.

21. Bouchet, *Le Cheval à Paris,* 92.

22. Lewis A. Merillat and Delwin M. Campbell, *Veterinary Military History of the United States,* 2 vols. (Chicago: Veterinary Magazine Corp., 1935), 1:701; G. Terry Sharrer, "The Great Glanders Epizootic, 1861–66," *Agricultural History* 69 (Winter 1995): 83. There were also very few equine deaths due to combat in the Civil War.

23. The best work on this topic is Susan D. Jones, *Valuing Animals: Veterinarians and Their Patients in Modern America* (Baltimore: Johns Hopkins University Press, 2003).

24. Ibid., 11.

25. *Seventh Census of the United States* (1850), 1:lxxvi. This was the first census to count veterinarians.

26. *AVR* 23 (July 1899): 318.

27. Department of the Interior, Census Office, *Eleventh Census, 1890,* "Population, Part 2," 304–5.

28. B. W. Bierer, *A Short History of Veterinary Medicine in America* (East Lansing: Michigan State University Press, 1955), 28.

29. Ezra Pater, *The Fortune Teller and Experienced Farrier in Two Parts* (Exeter, N.H., 1794). The Boston veterinarian adopted the title without training, since no woman graduated from a veterinary school before 1897.

30. J. Carver, Veterinary Surgeon, *Veterinary Science, Important to the Physician, the Student, and the Gentleman of Philadelphia* (Philadelphia, 1817), 42–3, is the earliest work including an attack on folk practitioners that we found.

31. "Pink Eye Epidemic among Horses," *Expressmen's Monthly* 6 (Nov. 1881); A. F. Liutard, "Latent Glanders," *AVR* 3 (Feb. 1880): 418–23. Jones, *Valuing Animals,* suggests that these traditional remedies did not reflect an innate sadism but rather were used to comply with the expectations of animal owners.

32. William Youatt, *The Horse with a Treatise on Draught and a Copious Index* (London: Baldwin & Craddock, 1861), 121–31, 175–8, 208, 324.

33. Edward Mayhew, *Illustrated Horse Doctor* (Philadelphia, 1861), 40, 132–46. Edward Mayhew, *Illustrated Horse Management* (Philadelphia, 1867), 347, argued that "an occasional quart of whiskey is good for a horse."

34. *Expressman's Monthly* 6 (Nov. 1881): 261.

35. William Rosser, "Corollary Development of the Professions of Veterinary Medicine and Human Medicine in the United States," *Veterinary Heritage* 14 (Mar. 1891): 12–16; Merillat and Campbell, *Veterinary Military History,* 1:67–81.

36. Merillat and Campbell, *Veterinary Military History,* 1:148–51; Sharrer, "The Great Glanders Epizootic," 8, 95: Rosser, "Veterinary Medicine and Human Medicine," 17.

37. Sharrer, "The Great Glanders Epizootic," 83; Charles W. Ramsdell, "Lee's Horse Supply," *American Historical Review* 35 (July 1930): 758–77.

38. Sharrer, "The Great Glanders Epizootic," 92; James Law, "Diseases and Management of Animals," in L. H. Bailey, ed., *Cyclopedia of American Agriculture: A Popular Survey of Agricultural Conditions, Practices and Ideals in the United States and Canada,* 4 vols. (New York: Macmillan Co., 1908), 3:123–8; G. W. Butler, "Hygiene and Its Relation to the Horse," *Annual Report of the State Board of Agriculture of Ohio* (1893), 426–30.

39. John R. Page, M.D., *The History, Symptoms, Diagnosis and Treatment of Glanders and Farcy in Horses* (Charlottesville, Va.: Chronicle Printing Press, 1876); Sharrer, "The Great Glanders Epizootic," 80. Sharrer points out that Page may have been aware of earlier European studies. See also H. J Detmers, "Glanders," *Annual Report of the State Board of Agriculture of Ohio* (1879), 532–74.

40. N. Kinsman, "Glanders in Horses," *Annual Report of the State Board of Agriculture of Ohio* (1886), 106–9.

41. *Annual Report of the Ohio State Livestock Commission* (1893), 55. This belief should be regarded with some skepticism. The elimination of glanders was fifty years off.

42. Sharrer, "The Great Glanders Epizootic," 95–6; Lise Wilkinson, *Animals and Disease: An Introduction to the History of Comparative Medicine* (Cambridge: Cambridge University Press, 1992), 109–14; Bierer, *Short History of Veterinary Medicine,* 53. The best histories of the germ theory in its relationship to animals are Michael Worboys's article, "Germ Theories of Disease and British Veterinary Medicine," *Medical History* 35 (July 1991): 308–27, and his book *Spreading Germs: Disease Theories and Medical Practice in Britain, 1865–1900* (Cambridge: Cambridge University Press, 2000). Worboys notes great variations in interpretations of the germ theory. Chapter 1 notes its early acceptance by veterinarians, although older and younger practitioners understood it differently.

43. Adoniram Judson, "History and Course of the Epizootic among Horses on the North American Continent," *American Public Health Association Reports* 1 (1873): 88–109; "The Epizooty Again," *Journal of Comparative Medicine and Surgery* 3 (Sept. 1881): 294.

44. Billings, *Relation of Animal Diseases to Public Health,* 174.

45. "Exposure of a Glandered Horse in a Public Market—Conviction—Sentence," *AVR* 1 (Feb. 1878): 304–9; Billings, *Relation of Animal Diseases to Public Health,* 174–81; W. L. Williams, "The Influence of Climate and Other Environments on the Distribution and Character of Disease," *AVR* 18 (Nov. 1894): 543, 547.

46. "Editorial: The Blackmailing of Veterinarians by Horse Dealers," *AVR* 22 (Nov. 1898): 517–9; Billings, *Relation of Animal Diseases to Public Health,* 174–81.

47. Worboys, "Germ Theories of Disease," 308; Jones, *Valuing Animals,* 28; *AVR* 2 (Aug. 1880): 170–6; *AVR* 9 (Jan. 1886): 2.

48. Merillat and Campbell, *Veterinary Military History,* 1:299; Jones, *Valuing Animals,* 32; "Report on Glanders," in Connecticut Commissioner on Domestic Animals, *Biennial Report* (1912), 23.

49. Merillat and Campbell, *Veterinary Military History,* 1:289.

50. Olof Schwartzkopf, "The Horse as Producer of Anti-toxins," *Journal of Comparative*

Medicine and Surgery 17 (Jan. 1896): 13–17. Horses still provide an important source of pharmaceuticals, most notably estrogen, which is derived from the urine of pregnant mares. See Harold Barclay, *The Role of the Horse in Man's Culture* (London: J. A. Allen, 1980), 161. Barclay wrote that about one hundred specialized farms in Ontario were providing the substance.

51. Charles Francis Adams Jr., "The Importance of Veterinary Science," *Harvard Graduate's Magazine* 3 (Dec. 1894): 188–92; Sharrer, "The Great Glanders Epizootic," 96; Bierer, *Short History of Veterinary Medicine*, 95; Jones, *Valuing Animals*, 7.

52. Worboys, "Germ Theories of Disease," 308–27; Adams, "The Importance of Veterinary Science," 188–92; F. B. Carleton, "Homeopathy Applied to Veterinary Medicine," *AVR* 18 (Aug. 1894): 333–6.

53. *Thirteenth Census of the United States, Taken in the Year 1910*, vol. 4, *Population: Occupation Statistics* (Washington, D.C.: Government Printing Office, 1912), 122. Jones, *Valuing Animals*, 13, points out that one of the rationales for excluding women was that castration, hardly appropriate work for "ladies," was an important part of many veterinary practices.

54. E. L Volgenu, "The Future of the Veterinary Profession," *AVR* 21 (June 1897): 183–6.

Eight • The Decline and Persistence of the Urban Horse

1. Dolores Greenberg, "Reassessing the Power Patterns of the Industrial Revolution: An Anglo-American Comparison," *American Historical Review* 87 (Dec. 1982): 1237–61, and "Energy, Power, and Perceptions of Social Change in the Early Nineteenth Century," *American Historical Review* 95 (June 1990): 693–714; Louis C. Hunter, *A History of Industrial Power in the United States, 1780–1930*, vol. 1, *Waterpower in the Century of the Steam Engine* (Charlottesville: University Press of Virginia, 1979), 111; Bruce Laurie and Mark Schmitz, "Manufacture and Productivity: The Making of an Industrial Base, Philadelphia, 1850–1880," in Theodore Hershberg, ed., *Philadelphia: Work, Space, Family, and Group Experience in the Nineteenth Century* (New York: Oxford University Press, 1981), 43–92.

2. "Motive Power," Transactions of the American Institute, *Proceedings of the Polytechnic Association* (1860), 53–9; "Horse Power Machine," *Scientific American* 11 (Sept. 17, 1864): 176; Jennifer Tann, "Horsepower, 1780–1880," in F. M. L. Thompson, ed., *Horses in European Economic History: A Preliminary Canter* (Reading: British Agricultural History Society, 1983), 26–30.

3. Louis C. Hunter and Lynwood Bryant, *A History of Industrial Power in the United States, 1780–1930*, vol. 3, *The Transmission of Power* (Cambridge: MIT Press, 1991), 49–51; Sarah Bradford Landau and Carl W. Condit, *Rise of the New York Skyscraper, 1865–1913* (New Haven: Yale University Press, 1996), 39; "Catalogue E," in *Clyde Steam and Electric Hoisting Engines and Derricks* (Duluth: Clyde Iron Works, 1912), 118–21. The builders of the New York subway loaded mules into their tunnels in 1900 when they started construction. The mules never saw the light of day while they worked in the tunnels and continued working underground until the subway was finished four years later. *New York Times* [hereafter cited as *NYT*], Feb. 25, 2003. Mules were preferred for tunneling, probably because they were generally shorter than horses.

4. Kevin J. Crisman and Arthur B. Cohn, *When Horses Walked on Water: Horse-powered Ferries in Nineteenth-Century America* (Washington, D.C.: Smithsonian Institution Press, 2002).

5. *Census of Massachusetts* (1895), vol. 5, *Manufacturing,* 341; Allied Machinery Co., *Hoisting Machines* [Manufactured by Clyde Iron Works], (New York, 1920); Clyde Iron Works, *Clyde Hoisting Engines and Derricks;* United States Wind Engine & Pump Co., *General Catalogue No. 15* (Batavia, Ill., 1910, 1919). See also Joel A. Tarr, "A Note on the Horse as an Urban Power Source," *Journal of Urban History* 25 (Mar. 1999): 434–8. Horses also persisted as power sources in long-distance canal travel. Shippers on the Erie Canal (completed in 1825) had attempted to switch to steam power as early as the 1850s, but horse use persisted primarily because of its flexibility. As late as 1890 (sixty years after the first American railroad), relays of horses pulled 1,763 towboats over 381 miles, moving 3,673,554 tons of freight.

6. Barbara Young Welke, *Recasting American Liberty: Gender, Race, Law, and the Railroad Revolution, 1865–1920* (New York: Cambridge University Press, 2001), x, 6–7.

7. Francis V. Greene, "An Account of Some Observations of Street Traffic," *American Society of Civil Engineers Transactions* 15 (1886): 123–38; Asha Weinstein, "The Congestion Evil: Perceptions of Traffic Congestion in Boston in the 1890s and 1920s," Ph.D. diss., University of California–Berkeley, 2002, 147; Clay McShane, *Down the Asphalt Path: American Cities and the Automobile* (New York: Columbia University Press, 1995), 190–200.

8. *NYT,* Nov. 13, 1885.

9. New York City Board of Health, *Annual Report, 1890,* 176–7; McShane, *Down the Asphalt Path,* 175–6; Timothy F. Kruse, "Teamsters in the Gilded Age," M.A. thesis, University of Wisconsin–Madison, 1970, 56. See also Roger Cooter and Bill Luckin, eds., *Accidents in History* (Amsterdam: Rodopi, 1997).

10. Welke, *Recasting American Liberty,* 70.

11. For increasing litigation, see Randolph E. Bergstrom, *Courting Danger: Injury and Law in New York City, 1870–1910* (Ithaca: Cornell University Press, 1992), 37. Passengers and bystanders were involved in a bewildering array of accidents. Pedestrians posed particular problems because they were prone, as the owner of one streetcar company noted, to "nervousness, carelessness and often times utter recklessness," placing the blame on the human rather than the technology. See C. A. Richards, "Accidents, Their Cause, Prevention and Settlement," *Street Railway Journal* 3 (Nov. 1886): 4–7. Other corporations with large herds put internal procedures in place to reduce the effects of litigation. In 1902, for example, Wells Fargo laid out elaborate accident report procedures for its drivers. See Wells Fargo & Co. Express, *Rules and Instructions* (San Francisco, 1902), n.p.

12. Henry Bixby Hemenway, M.D., *Essentials of Veterinary Law* (Chicago: T. H. Flood & Co., 1916), 33. Shibasabru Kitasato demonstrated the adverse effects on human health of flies bred in manure in a series of experiments in 1899. See Kenneth B. Haas, "The Trouble with Horse Manure," *Veterinary Heritage* 21 (Dec. 1998): 37; Naomi Rogers, "Germs with Legs: Flies, Disease, and the New Public Health," *Bulletin of the History of Medicine* 63 (1989): 599–617; Carolyn Hodak, "Les Animaux dans la cité: Pour une histoire urbaine," trans. Yvette Chin, *Génèses* 37 (Dec. 1999): 156–69; and Hemenway, *Essentials of Veterinary Law,* 33.

13. *Street Railway Journal* [hereafter cited as *SRJ*], 4 (Apr. 1888): 49; *NYT*, May 28–June 3, 1887; Tim Gilfoyle, "The Moral Origins of Political Surveillance: The Preventative Society in New York City, 1867–1918," *American Quarterly* 38 (Fall 1986): 637–52; Bernard Oreste Unti, "The Quality of Mercy: Organized Animal Protection in the United States, 1866–1930," Ph.D. diss., American University, 2002, 187. It is not clear what the humane societies thought the fate of the newly unemployed horses would be.

14. See "How Horse Cars Are Run," *The Stable* 2 (Mar. 1887): 28.

15. "Growth of City Traffic," *SRJ* 3 (Dec. 1887): 1047; Social Statistics of Cities, *1880 Census*, vols. 17 & 18; Charles H. Cooley, "Statistics of Street Railway Transportation," *Eleventh Census of the United States*, vol. 22 (1890): 682; *Metropolitan Street Railway Annual Reports, 1880 and 1890*, in *Annual Reports of Massachusetts Railroad Commission* (Boston: Wright & Potter, 1880, 1890); *Annual Reports of the New York State Engineer* (Albany: Weed, Parsons & Co., 1880, 1885). The Second Avenue Street Railway in New York depreciated its horses at 7% in 1880 but at 16% in 1885, suggesting a heavier workload.

16. Joel A. Tarr, "From City to Suburb: The 'Moral' Influence of Transportation Technology," in Tarr, *The Search for the Ultimate Sink: Urban Pollution in Historical Perspective* (Akron, Ohio: University of Akron Press, 1996), 309–22; Jon C. Teaford, *City and Suburb: The Political Fragmentation of Metropolitan America, 1850–1920* (Baltimore: Johns Hopkins University Press, 1979).

17. For a full account of the controversy over the elevated, see McShane, *Down the Asphalt Path*, 74–5. "The Horse vs. the Locomotive," *Manufacturer and Builder* 7 (July 1875): 188, editorialized that the horse was superior to the dummy steam engine for urban use.

18. *SRJ* 4 (Mar. 1888): 97; Craig Semsel, "The Mechanization of Pittsburgh Street Railways, 1886–1897," *Pittsburgh History* 77 (1994): 54–67; George W. Hilton, *The Cable Car in America* (Berkeley, Calif.: Howell-North Books, 1971).

19. McShane, *Down the Asphalt Path*, 28.

20. Cooley, "Statistics of Street Railway Transportation," 681–5; Clay McShane, *Technology and Reform: Street Railways and the Growth of a City* (Madison: State Historical Society of Wisconsin, 1974), ch. 2; U.S. Census, *Report on Street and Electric Railways, 1902* (Washington, D.C.: GPO, 1905), 7–12; Herbert N. Casson, "The Horse Cost of Living," *Munsey's Magazine* 48 (Mar. 1913): 1997; Tarr, *Search for the Ultimate Sink*, 328–30.

21. This situation contrasted with that in Europe, where many large cities banned overhead electric traction lines. See John P. McKay, *Tramways and Trolleys: The Rise of Urban Mass Transport in Europe* (Princeton: Princeton University Press, 1976), 84–95.

22. *Additional Burdens on Street Railway Companies, Arguments of Henry Whitney and Prentiss Cummings before the Committee on Cities and the Committee on Taxation of the Massachusetts Legislature* (Boston: Press of Samuel Usher, 1891); Louis P. Hager, ed., *History of the West End Street Railway* (Boston: Privately printed, 1892), 11–8; Frank J. Sprague, "The Future of the Electric Railway," *Forum* 12 (Sept. 1891): 120–30; Leslie F. Blanchard, *The Street Railway Era in Seattle: A Chronicle of Six Decades* (Forty Fort, Pa.: H. E. Cox, 1968), 4–8; Robert M. Fogelson, *The Fragmented Metropolis: Los Angeles, 1850–1930* (Cambridge: Harvard University Press, 1967), 35–9; Harold C. Passer, *The Electrical Manufacturers, 1875–1900* (Cambridge: Harvard University Press, 1953), 218–47.

23. Tarr, *Search for the Ultimate Sink*, 329–30.

24. A few omnibuses had remained on elite streets whose residents had successfully opposed the introduction of overhead wires or on terminal-hotel runs. These mechanized rapidly after the introduction of imported DeDion Bouton buses on Fifth Avenue in New York City in 1906. Horse-drawn cabs were another aspect of the urban scene that almost completely disappeared except as a tourist attraction. The first attempt at mass-producing cars in the United States was the Electric Vehicle Company's attempt to produce electric cabs for New York and other cities in 1899. While these failed, horse-drawn cabs completely disappeared during the jitney craze of 1914–15, when Model T owners organized unlicensed "gypsy" cablike service in virtually every city in the country. Thus, the transition took fifteen years. McShane, *Down the Asphalt Path*, 34, 195; Clay McShane, *The Automobile: A Chronology of Its Antecedents, Development, and Impacts* (Westport, Conn.: Greenwood Press, 1997), 34–6; David A. Kirsch, *The Electric Vehicle and the Burden of History* (New Brunswick, N.J.: Rutgers University Press, 2000), 29–84; "Editorial," *International Teamster* 6 (Sept. 1905): 5; Ross Eckert and George W. Hilton, "The Jitneys," *Journal of Law and Economics* 15 (Oct. 1972): 72.

25. Paul Barrett, *The Automobile and Urban Transit: The Formation of Public Policy in Chicago, 1900–1930* (Philadelphia: Temple University Press, 1983), 53–4.

26. Don H. Berkebile, ed., *Horse-drawn Commercial Vehicles* (New York: Dover, 1989), 11, 75. In 1894, a New York manufacturer built a thirty-two-foot long, seven-ton truck to move heavy machinery.

27. Kirsch, *The Electric Vehicle*, 136–7; W. P. Hedden, *How Great Cities Are Fed* (Boston: D. C. Heath, 1929), 184–5. We thank Melanie Dupuis for calling this book to our attention.

28. "The Passing of the Horse," *NYT*, June 10, 1894, Feb. 13, 1896; *Expressmen's Monthly* 20 (Sept. 15, 1895): 211; Robert H. Thurston, "The Automobile in Traction," *Review of Reviews* 21 (Feb. 1900): 225; James J. Flink, *America Adopts the Automobile, 1895–1910* (Cambridge: MIT Press, 1970), 21.

29. Kirsch, *The Electric Vehicle*, 136–44.

30. Ibid., 137–45; Gijs Mom, *The Electric Vehicle: Technology and Expectations in the Automobile Age* (Baltimore: Johns Hopkins University Press, 2004), 7, 206. For road conditions and improvements, see Bruce E. Seely, *Building the American Highway System: Engineers as Policy Makers* (Philadelphia: Temple University Press, 1987).

31. Mom, *The Electric Vehicle*, 7.

32. Fulton Green, *National Survey of the Economic Status of the Horse* (Detroit, 1920), 52–64. See also Alan L. Olmstead and Paul W. Rhode, "The Agricultural Mechanization Controversy of the Interwar Years," *Agricultural History* 68 (Summer 1994): 41–43.

33. A Philadelphia ice concern used one thousand horses and twenty-five motor trucks in its business, while a Detroit creamery used five hundred horses and one hundred trucks. In both cases the trucks served long-distance requirements. See Green, *Economic Status of the Horse* (Detroit, 1920), 60–3.

34. In Pittsburgh, the number of horse-drawn vehicles (almost certainly close to 100% goods carrying) in the central business district during a twelve-hour period declined from 8,370 in 1917 to 7,545 in 1920, while the number of motor trucks rose from 6,432 to 20,554 and the number of automobiles from 9,811 to 32,416. See Joel A. Tarr, *Transportation In-*

novation and Spatial Change in Pittsburgh, 1850–1934 (Chicago: Public Works Historical Society, 1978), 26–7.

35. Gijs P. A. Mom and David Kirsch, "Horses, Electric Trucks and the Motorization of American Cities, 1900–1925," *Technology and Culture* 42 (July 2001): 495–7; City of Pittsburgh, *City of Pittsburgh and Its Horses* (Pittsburgh, 1930), 3–24. As late as 1930 the City of Pittsburgh owned three hundred horses, mostly draft animals held by the Public Works Department. The municipality claimed that all of them could understand traffic lights and that the local humane society had commended its herd management. Joel Tarr remembers horses pulling garbage trucks, peddler's wagons, and scrap wagons on Jersey City streets in the 1940s. One group of fruit and vegetable peddlers still plied city streets in horse-pulled wagons in 1989, the "arabbers" of Baltimore. Roland L. Freeman, *The Arabbers of Baltimore* (Centreville, Md.: Tidewater Publishers, 1989). Arabbers still work on Baltimore streets today. See the Arabber Preservation Home Page, www.baltimoremd.com/arabber (visited Jan. 5, 2007). The Horse Association of America struggled to preserve the horse's role in the city; for instance, it opposed automobile parking on downtown streets because it impeded horse deliveries. See Olmstead and Rhode, "The Agricultural Mechanization Controversy," 40–42.

36. W. R Metz, "Cost of Upkeep of Horse-drawn Vehicles against Electric Vehicles," *American Society of Mechanical Engineers Proceedings* (1913): 129–38; John W. Linehan and Edward Cogswell, eds., *The Driving Clubs of Greater Boston* (Boston: Atlantic Printing Co., 1914), 24; Green, *Economic Status of the Horse,* 52–3.

37. Green, *Economic Status of the Horse,* 66–7.

38. Hodak, "Les Animaux," 166–7. PartyPop.com, www.partypop.com/search.cfm?q=horse (visited Aug. 20, 2005), presented a 2003 listing of different states and cities where one could rent a horse and carriage. They advertised the romance of a carriage ride with the following paragraph: "A horse drawn carriage ride in the park under the stars is one of the most romantic moments two people in love can share. A horse and buggy ride just creates that magical feeling that only an open ride in a horse and carriage can create. What is it about the Horse And Buggie [*sic*] Ride that is so special? Perhaps because it's a throwback to another era, to times that we only read about in storybooks and fantasise [*sic*] about." Renting a horse in Central Park costs a whopping fifty dollars an hour. The same site lists twenty-three vendors of hayrides in or near big cities, as well as sleigh rides where the climate is appropriate.

39. Elizabeth Atwood Lawrence, *Hoofbeats and Society: Studies of Human-Horse Interactions* (Bloomington: Indiana University Press, 1985), 116–73, has a description of the contemporary police. Mounted police insist on geldings, not mares, and prefer very large, dark-colored horses.

Epilogue

1. *New York Times,* July 24, 1881.

2. We have argued elsewhere that the horse was a technology that grew increasingly inefficient as the modern, networked city grew around it. In this sense, the horse became what historian of technology Thomas Parke Hughes calls a reverse salient, or a lagging el-

ement in a technological system holding back the development of the network. See Clay McShane and Joel A. Tarr, "The Centrality of the Horse in the Nineteenth-Century American City," in Ray A. Mohl, ed., *The Making of Urban America,* 2d ed. (Wilmington, Del.: Scholarly Resources, 1997), 105–30; Thomas Parke Hughes, *Networks of Power: Electrification in Western Society, 1880–1930* (Baltimore: Johns Hopkins University Press, 1983), 14–5; and "The Evolution of Large Technological Systems," in W. E. Bijker, T. P. Hughes, and T. J. Pinch, eds., *The Social Construction of Technological Systems: New Directions in the Sociology and History of Technology* (Cambridge: MIT Press, 1987), 51–82.

3. There are statues of horses in many cities. In central Boston, for instance, there are ten statues of horses, mostly carrying warriors. Of these, only one, outside the entrance of Massachusetts General Hospital, is a draft horse. These data are based on a count by students from Clay McShane's History of Boston class. Such statues are common in other cities also. See Annabelle Sabloff, *Reordering the Natural World: Humans and Animals in the City* (Toronto: University of Toronto Press, 2001), for a case study of Toronto.

4. Based on a subject search of the electronic catalog Worldcat.

5. Joel Tarr, *The Impact of Transportation Innovation on Changing Spatial Patterns: Pittsburgh, 1850–1934* (Chicago: Public Works Historical Society, 1978); Clay McShane, *Down the Asphalt Path: American Cities and the Automobile* (New York: Columbia University Press, 1994). Rudi Volti, *Cars and Culture: The Life Story of a Technology* (Baltimore: Johns Hopkins University Press, 2006), is the best recent survey of automobile history.

6. *Thirteenth Census of the United States, Taken in the Year 1910,* vol. 3, *Population* (Washington, D.C.: Government Printing Office, 1915), 201; *Twelfth Census of the United States, Taken in the Year 1900,* vol. 5, *Part 1, Agriculture* (Washington, D.C.: Government Printing Office, 1902), 513; New York State Department of Motor Vehicles, *Statistics* (www.nydmv.state.ny.us/statistics/regin5, visited July 30, 2006). Technically, these are figures for New York County, which is quite nearly coterminous with Manhattan. Horses were counted where they were stabled and cars where they were kept at night. At both times, especially, we suspect, the latter, daytime populations were much higher. By way of contrast, the United States as a whole had almost as many motor vehicles as people in 2000, and Los Angeles had 1.3 persons per motor vehicle.

7. Tertius Chandler, *Four Thousand Years of Urban Growth, a Historical Census* (Lewiston, N.Y.: St. David's University Press, 1987).

8. "Some Advantages of the Automobile," *Horseless Age* 10 (Oct. 8, 1902): 377.

9. *Sprawl* is a loaded word. For a lengthy discussion of its origins, usage, and connotation, see the postings on the electronic discussion group H-Urban, best accessed by going to its home page at www.h-net.org/urban/ and entering the word "sprawl" in the search function. See Owen D. Gutfreund, *Twentieth-Century Sprawl: Highways and the Reshaping of the American Landscape* (New York: Oxford University Press, 2004), and Robert Bruegman, *Sprawl: A Compact History* (Chicago: University of Chicago Press, 2005).

10. Joel A. Tarr, "From City to Suburb: The 'Moral' Implications of Transportation Technology," in Tarr, *The Search for the Ultimate Sink: Urban Pollution in Historical Perspective* (Akron, Ohio: University of Akron Press, 1996), 309–22.

INDEX

Page numbers in italics refer to figures and tables. The gallery of photographs follows text page 56 *(photo)*; tables are indicated by the letter *t*.

Milton Keynes UK
Ingram Content Group UK Ltd.
UKHW010627250424
441709UK00001B/49